WITHDRAWN BY THE
UNIVERSITY OF MICHIGAN

Peter von Buelow

Genetically Engineered Architecture

Peter von Buelow

Genetically Engineered Architecture

Design Exploration with Evolutionary Computation

VDM Verlag Dr. Müller

Bibliographic information by the German National Library: The German National Library lists this publication at the German National Bibliography; detailed bibliographic information is available on the Internet at http://dnb.d-nb.de.

This works including all its parts is protected by copyright. Any utilization falling outside the narrow scope of the German Copyright Act without prior consent of the publishing house is prohibited and may be subject to prosecution. This applies especially to duplication, translations, microfilming and storage and processing in electronic systems.

Any brand names and product names mentioned in this book are subject to trademark, brand or patent protection and are trademarks or registered trademarks of their respective holders. The use of brand names, product names, common names, trade names, product descriptions etc. even without a particular marking in this works is in no way to be construed to mean that such names may be regarded as unrestricted in respect of trademark and brand protection legislation and could thus be used by anyone.

Copyright © 2007 VDM Verlag Dr. Müller e. K. and licensors
All rights reserved. Saarbrücken 2007
Contact: info@vdm-verlag.de
Cover image: www.purestockx.com
Publisher: VDM Verlag Dr. Müller e. K., Dudweiler Landstr. 125 a, 66123 Saarbrücken, Germany
Produced by: Lightning Source Inc., La Vergne, Tennessee/USA
           Lightning Source UK Ltd., Milton Keynes, UK

Bibliografische Information der Deutschen Nationalbibliothek: Die Deutsche Nationalbibliothek verzeichnet diese Publikation in der Deutschen Nationalbibliografie; detaillierte bibliografische Daten sind im Internet über http://dnb.d-nb.de abrufbar.

Das Werk ist einschließlich aller seiner Teile urheberrechtlich geschützt. Jede Verwertung außerhalb der engen Grenzen des Urheberrechtsgesetzes ist ohne Zustimmung des Verlages unzulässig und strafbar. Das gilt insbesondere für Vervielfältigungen, Übersetzungen, Mikroverfilmungen und die Einspeicherung und Verarbeitung in elektronischen Systemen.

Alle in diesem Buch genannten Marken und Produktnamen unterliegen warenzeichen-, marken- oder patentrechtlichem Schutz bzw. sind Warenzeichen oder eingetragene Warenzeichen der jeweiligen Inhaber. Die Wiedergabe von Marken, Produktnamen, Gebrauchsnamen, Handelsnamen, Warenbezeichnungen u.s.w. in diesem Werk berechtigt auch ohne besondere Kennzeichnung nicht zu der Annahme, dass solche Namen im Sinne der Warenzeichen- und Markenschutzgesetzgebung als frei zu betrachten wären und daher von jedermann benutzt werden dürften.

Copyright © 2007 VDM Verlag Dr. Müller e. K. und Lizenzgeber
Alle Rechte vorbehalten. Saarbrücken 2007
Kontakt: info@vdm-verlag.de
Coverbild: www.purestockx.com
Verlag: VDM Verlag Dr. Müller e. K., Dudweiler Landstr. 125 a, 66123 Saarbrücken, Deutschland
Herstellung: Lightning Source Inc., La Vergne, Tennessee/USA
           Lightning Source UK Ltd., Milton Keynes, UK

ISBN: 978-3-8364-4721-8

Table of Contents

1 DESIGN IN ARCHITECTURAL ENGINEERING 1

1.1 Definition of Design ... 2
   1.1.1 Trichotomy of Design ... 2
   1.1.2 Practice of Design ... 5

1.2 Study of Design ... 9
   1.2.1 Theory and Practice of Design ... 9
   1.2.2 Design Models ... 13
   1.2.3 Design Mechanisms ... 23

1.3 Tools for Design ... 32
   1.3.1 Non-computational Design Tools ... 33
   1.3.2 Computational Analysis *versus* Design ... 41
   1.3.3 Computational Design Tools ... 46
   1.3.4 The IGDT Design Tool ... 63

2 THE INTELLIGENT GENETIC DESIGN TOOL 69

2.1 Constructing Genetic Tools ... 69
   2.1.1 Design Objectives ... 69
   2.1.2 Encoding Techniques ... 70
   2.1.3 Search and Exploration ... 76

2.2 Implementation of the IGDT ... 85
   2.2.1 Defining Problem Parameters ... 85
   2.2.2 Topology Search ... 86
   2.2.3 Geometry Search ... 88
   2.2.4 Running the IGDT ... 92

3 EXAMPLES AND RESULTS 95

   3.1.1 Flat Deck Bridge ... 96
   3.1.2 Problem Description and Setup ... 96
   3.1.3 Use of the IGDT ... 97
   3.1.4 Comparison of Results ... 104
   3.1.5 Conclusions ... 108

3.2 Arch Truss ... 108
   3.2.1 Problem Description and Setup ... 108
   3.2.2 Use of the IGDT ... 110
   3.2.3 Comparison of Results ... 112
   3.2.4 Conclusions ... 113

3.3 Cantilever Truss ... 113
   3.3.1 Problem Description and Setup ... 113
   3.3.2 Use of the IGDT ... 114
   3.3.3 Comparison of Results ... 117
   3.3.4 Conclusions ... 119

3.4 High Speed Gantry 119
3.4.1 Problem Description and Setup 119
3.4.2 Use of the IGDT 121
3.4.3 Comparison of Results 123
3.4.4 Conclusions 125

3.5 Interactive Design 126
3.5.1 Problem Description and Setup 126
3.5.2 Use of the IGDT 127
3.5.3 Comparison of Results 129
3.5.4 Conclusions 130

4 CONCLUSION   131

4.1 Summary 131
4.1.1 Aspects of Design 131
4.1.2 Aspects of GAs and the IGDT 131

4.2 Results and Recommendations 132
4.2.1 Applications 132
4.2.2 Current Limitations 133
4.2.3 Further Development 135

4.3 Closing Remarks 138

5 REFERENCE LIST   141

APPENDIX A: EXAMPLE INPUT/OUTPUT FILES FOR THE IGDT   150

Example Input Data File 150

Example Geometry File for Progenitor 151

Example Output Text File for one Geometry 152

APPENDIX B: GRAPHIC DEPICTION OF THE GEOMETRY CHC   153

APPENDIX C: GRAPHIC DEPICTION OF THE TOPOLOGY ES   154

APPENDIX D: ANALYSIS ASSESSMENT   155

Comparison of Member Forces 155

Comparison of Section Areas 155

1 Design in Architectural Engineering

This book describes a new direction in the use of computational aids to the age old process of design. Whereas, in recent decades millions of lines of programming code have been directed at providing designers with help in their activity, almost all of this work has been limited to the technical or knowledge based side of their activity, with almost no successful attempts to provide aid to the creative aspects of design as well. This book presents a new approach which is intended to stimulate the designer's own creativity as well as offering technical assistance. By being sensitive to the creative aspects of the design process, the proposed design aid can be employed with success in earlier, conceptual phases of design, where many currently offered tools fail to provide useful assistance.

In discussing design tools, I draw a distinction between design and analysis which is often over looked by developers of computer programs. There have been over the years several, very sophisticated, analysis tools developed, which through their user friendly input and easy to interpret output have been applied to the area of design. But I would maintain, that there remains a fundamental distinction between the activities present in analysis versus design, and that the use of analysis tools for the activity of design is at best ineffective and at worst detrimental to the creative process. Even as machines have filtered into almost all aspects of the profession, many designers still maintain a highly skeptical attitude toward the value of computer assistance in conceptual design. This is true to the extent that many architecture schools deliberately make no use of computers in the fundamental design courses. It is not unusual that some design faculty members will discourage (to the point of forbidding) the use of computers in all levels of design. One might attribute such behavior to a type of rigid conservatism and inability to adapt to changing methods - a condition certain to fade as the older, non-computing generation of designers leave the profession. But again, on closer inspection, it is often younger, computer-experienced designers who hold this position.

In this book I have tried to isolate some of those aspects of computer usage in design which have given cause for so much concern. The problem, I believe, lies not so much in the media or the machine itself, as in the conceptual analogy most programmers and users have of computers. This is the analogy of the computer to our own brain, and programming as knowledge held or supplied to that brain. In this analogy the user/designer tends to be dominated by the "brain". The user takes on the roll of apprentice - posing questions to the master (computer), and respectfully awaiting the answer. I propose reversing the rolls, and letting the machine act as our apprentice, to make proposals which the master craftsman can critique, with the expectation that the craft will be improved. Those who have taught, know how stimulating provocative students can be to the development of one's own ideas. Computer tools used in conceptual design should, and can, exhibit this same quality of creative stimulation.

The program proposed in this book is based on the methods of Genetic Algorithms (GA's) - a numerical technique derived from evolutionary biology. The first chapter explores the context of design tools, and shows why the GA method is particularly well suited for the requirements of such a tool. The second chapter discusses in more detail the programmatic aspects of Genetic Algorithms and their analogy to evolutionary genetics. The third chapter explains the structuring of the design tool application which was developed - the Interactive Genetic Design Tool (IGDT). The fourth chapter offers examples applied to design problems and comparisons with other computer based tools. The fifth chapter contains a discussion of results and recommendations for applications of similar techniques in other areas of design.

The program itself is comprised of about 21,600 lines of ANSI-C code (approximately 400 pages). There are 176 routines compiled into 5 different executables that are run on a parallel LINUX cluster using PVM. The current cluster of 100 CPU's contains a mix of Intel Pentium III and IV class processors. Most of the examples in Chapter 4 were run on between 10 and 30 machines for times ranging from 1 to 15 hours depending on the number of load cases and nodes in the solutions. A table showing run times and problem parameters is shown in Chapter 5. Due to the size, a complete listing of the code is not provided. The parts of the header file which define parameters and list all of the routines are provided in Appendix A and D along with the input and output files. Appendix E also includes an assessment of the FEA analysis code of the IGDT as compared to a commercial FEA code.

1.1 Definition of Design

1.1.1 Trichotomy of Design

Design is an activity that is found in all sectors of human endeavor. Whenever *purposeful* consideration is given to a problem, design is present. Design is also *goal oriented* in that it seeks the solution to a specific problem. But in speaking about a "designed solution", the general implication is, that it contains something in addition to a standard, goal oriented, purposeful solution. A designed solution is generally understood to contain some aspect that makes it non-standard. A designed solution is *creative*. Thus, in describing the design activity, these three attributes are in some way present:

- purposeful
- goal oriented
- creative

It is reasonable, that any aid to the design process must also respond to these aspects of the activity. Designers have traditionally used many different aids or tools to help them in their work. This book will look briefly at some of those aids as they provide a context for the consideration of new design tools. Specifically, in the area of computer based design tools, many programs have been written in recent years which respond well to the first two aspects of design, viz. purposeful and goal oriented, but in the area of the third aspect of design, the creative aspect, computer aids seem to flounder, and are often criticized as producing more hindrance than help. The criticism is not wholly unjustified, as an over infusion of technical knowledge in the early stages of design development can actually stifle the development of exploratory, creative consideration of the problem. This point is developed further in Section 1.2.2.2.

The Intelligent Genetic Design Tool (IGDT) developed in this book is intended to offer the designer purposeful, goal oriented, as well as creative support.

1.1.1.1 Aspects of Design

Theorists have proposed numerous approaches to design. None the less, as stated above, it is generally agreed that design as an activity is purposeful, goal oriented and creative.

Purposeful. There is always a purpose which initiates a design. This may seem self evident, but in the context of the heuristic tools used by an IGDT, it is perhaps well to formally make the point. Many interesting and useful solutions can be stumbled upon purely by chance, and an appropriate problem might be found to which the solution could be advantageously applied. This type of opportunism may be clever, but it is not design. Design begins only after a purpose has been established. Without a purpose it is impossible to establish goals.

Goal Oriented. Beyond having a purpose, design is a goal oriented activity. A specific goal is established which can be described by criteria. The goal usually involves finding the solution which best meets the criteria. In order to find the best solutions, a search is usually a major part of the design process. This is why optimization techniques are often used in computational design. Optimization is a search method which attempts to find the solution which best satisfies the goal. In optimization, goal criteria are often called objectives, and defined by objective functions. In the terminology of genetic design methods, the goal criteria are called fitness functions. The IGDT is similar to a search tool, but actually it is more. It is a design exploration tool. The difference between search and exploration is discussed in Section 1.2.2.3. Further, it is conceptually different from most optimization methods in that beyond being purposeful and goal oriented it also stimulates the designer's own creative thinking.

Creative. Design is a creative activity in that it seeks new solutions. If creativity is lacking, then either an existing solution, or else no solution, will be applied to a problem. If only existing solutions were used, the only source of progress would be happenstance. Even then, some creativity would be required to recognize a chance discovery as a better solution. Methods or tools which supply the user with one 'best' solution can actually work against creativity. By offering the designer only one solution, or even just one solution at a time, the implication is that the shown solution is the only one worth considering. The designer need only agree to the offered solution and the task is complete. As discussed in Section 1.2.2.2., design fixation, a common problem not just for novice designers, is aggravated by design tools which suggest only one solution for the problem. Such tools do little to promote creative speculation. An IGDT, on the other hand, by always supplying a palette of solutions, requires the designer to continually view different possibilities. In the process of considering arrays of solutions, creative thinking is stimulated, and the likelihood of finding a new or unexpected solution increases.

1.1.1.2 Definitions of Design

In order to understand the requirements of a design tool, it is helpful to first examine what designers do, or what is understood as the design activity. Between fields of endeavor as well as within the allied fields of architecture and engineering, design can apply to a wide range of activities. Nonetheless, the aspects of *purposeful, goal oriented* and *creative*, as described above, are useful in formulating an understanding of what the design activity involves. Everyone has had experience with design to one degree or another, in much the same way that everyone has had some experience with singing. But, just as with singing, there exist levels of ability and understanding, that distinguish

the activity from the art. In a general sense, the Oxford English Dictionary defines design, the verb as:
   I.   To mark out; to indicate.
   II. To plan out.
   III. To sketch; to form or fashion a work of art.
and as a noun design is defined as:
   I. A plan or a scheme contrived in the mind.
   II. The preliminary sketch of a work of art; the plan of a building or part of it.
(Onions, OED, 1968)

In order "to plan out", a problem solving process is implied. As a process, the design activity is *purposeful* and *goal oriented*. The goal is to solve the problem as defined. Also, as a problem solving process it is necessarily *creative*. In designing an artifact, one must create new solutions to the problems presented as design criteria. If old solutions are adequate to solve the problem criteria, then there is no need for design (or re-design) and the solution is seen as an application of the previous design.

Although design itself may be "contrived in the mind", Christopher Alexander, Berkeley professor and founder of the Center for Environmental Structure, points out in his book *Notes on the Synthesis of Form*, that "the ultimate objective of design is form." (Alexander, 1967, p.15) More specifically, Alexander sees design as the "fit" between form and context.

> The form is a part of the world over which we have control, and which we decide to shape while leaving the rest of the world as it is. The context is that part of the world which puts demands on this form; anything in the world that makes demands of the form is context. Fitness is a relation of mutual acceptability between these two. (Alexander, 1967, pp. 18-19)

For Alexander, good design is the "effortless contact or frictionless coexistence" of form and context. Bad design occurs if there is a "misfit". Interestingly, Alexander also points out that the "misfit" is much the more practical to describe than the "good fit".

> I should like to recommend that we should always expect to see the process of achieving good fit between two entities as a negative process of neutralizing the incongruities, or irritants, or forces, which cause misfit. (Alexander, 1967)

In Section 1.3.3.3., it is shown how an IGDT can work in just this way, by negating the misfit individuals from the population of solutions.

In order to achieve Alexander's "goodness of fit" for novel problems, requires new solutions that attain the desired goals. And the discovery of these new solutions is what requires the third element of design, creativity. Creativity is sometimes defined by researchers as a novel combination of old ideas. This sort of definition makes measurement of creativity somewhat easier by regarding the improbability of the combination of ideas used. But there is an obvious limitation to this definition in that no truly new ideas are ever used or expected. Margaret A. Boden, Professor of Philosophy and Psychology at the University of Sussex, has written much about creativity in relation to artificial intelligence. She describes creativity in terms of exploration.

> ... examples show that exploration often leads to novel ideas. Indeed, it often leads to ideas, such as new forms of harmonic modulation, that are normally called creative. In that sense, then, conceptual exploration is a form of creativity.
> (Boden, 1994)

This is a broader understanding of creativity. To find something new, one explores. One explores not just the realm of known ideas, but also untested, unexplored realms as well. It is this aspect of exploration, and hence creativity, that separates the IGDT developed in this book from other 'design tools' currently available. Although other tools exist which are *purposeful* and *goal oriented*, the third aspect of design, the *creativity*, has been missing. By centralizing exploration in a way that is still purposeful and goal oriented, the IGDT offers the designer a true aid to the conceptual design process that has been lacking in computer aided tools.

1.1.1.3 Examples of Design

The balance of this first chapter discusses, through examples, different means that have been applied in the past to aid designers. These are divided into two groups:

- design mechanisms
- design tools

Design mechanisms are discussed in Section 1.2. They are techniques that have been applied in the past to help designers overcome different hindrances to design, particularly in the area of creativity. Margaret Boden refers to these mechanisms as being applied in "transforming conceptual spaces." (Boden, 1994, p. 82). Her notion of "transforming" is similar to the idea of "paradigm shift" (Kuhn, 1962), developed by MIT Professor Emeritus for Linguistics and Philosophy, Thomas Kuhn. Boden translates Kuhn's "paradigm shift", which he applied to developing scientific theories, down to the scale of individual design problems. The mechanisms referred to here, are intended to help the designer see the relevance of the space being explored in solving the problem at hand. In this sense they have found application in a wide range of design problems. Since the IGDT is a design exploration tool, some of these traditional mechanisms can be used in conjunction with the IGDT to give the designer a new perspective of the design problem.

Design tools are discussed in Section 1.3. They are physical (or in the case of programs, coded) aids that assist the designer in the exploration of a problem. Many of these tools are the standard repertoire of the designer. The IGDT is not seen as necessarily a replacement for other successful tools, but can offer advantages in exploring areas inaccessible to traditional tool. As in the case of any new tool, the advantage lies in its ability to reach hitherto unexplored solutions in the possible design space. Examples of the application of the IGDT are given in Chapter 4.

1.1.2 Practice of Design

The Intelligent Genetic Design Tool (IGDT) described in this book, is intended for the field of architectural engineering. In this context further detail can be added to the understanding of design.

1.1.2.1 Design Meaning

Design assumes further meaning when applied to a specific discipline. In referring to engineering disciplines, Clive L. Dym, Professor and Director of the Center for Design Education at Harvey Mudd College and researcher in the area of engineering, artificial intelligence and design, writes regarding the design process:

> Engineering design is the systematic, intelligent generation and evaluation of specifications for artifacts whose form and function achieve stated objectives and satisfy specified constraints.
> (Dym, 1991)

In common with the three part definition of design given in Section 1.1.1., are the concepts of *purposeful*: "... intelligent generation and evaluation of specifications for artifacts...", and *goal oriented*: "... achieve stated objectives and satisfy specified constraints...".

In the opening session of the Conceptual Design of Structures, an international symposium of the International Association for Shell and Spatial Structures (IASS) in 1996 in Stuttgart, Germany, Professor Jörg Schlaich, Director of the Institute for Structural Design at the University of Stuttgart, defined structural engineering in this way:

> Structural engineering is much more than providing scientific proof for an already existing phenomena. ... Structural engineering expressed through conceptual design, means to combine knowledge with intuition, experience with fantasy and aims at inventing an efficient structure including a unique form. (Schlaich, 1996, p.16)

Here one can more clearly read the roll of creativity in design. Although *purposeful* and *goal oriented* might apply just as well to analysis, "fantasy", "inventing" and "unique form" describe the *creative* aspects of the activity which qualifies it as design. These are also the aspects of the IGDT that distinguish it from more traditional analysis tools.

In the past, one problem with computational tools for the field of architectural engineering has been the lack of distinction between design tools and analysis tools. As a result, it is common practice to use analysis tools for design. For example, one might use a finite element program to supply stress level information about a given form being considered. But generally such programs contribute little to the designer's "fantasy" and seldom suggest a "unique form". Such analysis tools can greatly enhance the designer's understanding of a given structure, but they are severely limited in being able to inspire the designer in finding new or unexpected solutions. Even programs that are able to manipulate form in terms of an optimization process, are typically analysis oriented. They analyze a set of criteria to find the best fitting form. The limitation here is that although such programs may be able to find the optimal form for the given criteria, the given criteria may not describe the best solution for the problem. Such programs can actually stifle creative design in that they only offer solutions for the problem as described. Offered the 'optimal' solution, the tendency on the part of the designer is to accept it without questioning what limitations might be implied by the way the problem criteria are stated. In submitting the criteria for analysis the designer may be able to formulate the "knowledge", but will the "intuition" be included? The program may contain a data base of design "experience", but will it enhance the designer's "fantasy"? If not it has some serious limitations as a *design* tool, and will provide little aid to the designer searching for Schlaich's "unique form".

1.1.2.2 Design Phases

As a process, design is usually divided into phases which describe a progression of activity which begins with a client's desire to build and continues through the life cycle of the completed structure to demanufacture. The distinction of design phases, although somewhat artificial, is generally understood by designers to delineate the various activities which they perform in the process of designing. *The Architect's Handbook of Professional Practice* divides design activity into the following phases (Cryer, 1994, p. 526):

- Schematic Design Phase
- Design Development Phase
- Construction Documents Phase

In this context, the focus of an IGDT lies in the earlier, Schematic Design Phase, before a final solution is chosen. Expanding the design phase headings somewhat helps to provide an understanding of where the IGDT is meant to *aid* the designer.

Schematic Design Phase
   analysis of project requirements
   diagram studies
   assembly of data
   schematic design studies and recommended solution
   schematic design plans
   sketches and study models
   general project description
   engineering system concepts

Design Development Phase
   refinement of project requirements
   formulation of civil engineering systems
   formulation of structural systems
   formulation of mechanical and electrical systems
   selection of major building materials
   preparation of Design Development documents

Construction Documents Phase
   development of major detail conditions
   diagram study of major mechanical and electrical systems
   diagram study of major civil and structural systems
   architectural working drawings and specifications
   engineering working drawings and specifications

As was earlier noted, most computer based design aids operate in the Design Development Phase, where a system has already been chosen and is refined through analysis. The IGDT on the other hand, is intended as an exploration tool, appropriate to the Schematic Design Phase of gathering data and schematic studies of different possibilities.

During the 1960's several researchers gave consideration to the programmatic description of the design process (Coyne, et al., 1990, p.4). Most included some variation of the three step cycle proposed by Morris Asimow (Asimow, 1962):

- analysis
- synthesis
- evaluation

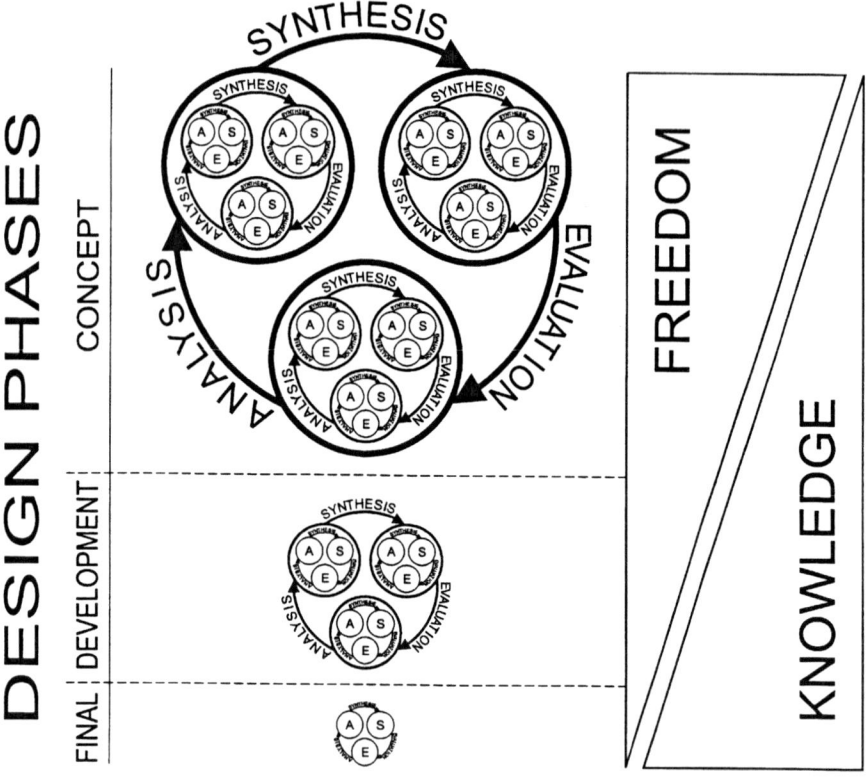

Figure 1.1. Nested and Interlinking Cycles in the Design Process.

Asimow's three step method is further discussed in Section 1.2.1.1. below. Some form of this cycle occurs, in a nested fashion, within each of the design phases. The process becomes a series of inter-linking cyclic shells in which the introduction of new knowledge or new events can cause regression to a previous point from any point in the process. Figure 1.1 depicts the interlinking of these three phases. Note that because less is known about the problem in the early phases, there is a greater likelihood, as well as need, for iteration. With each decision made in the process, both more knowledge is gained, and more constraints are imposed which limit the designer's freedom to consider multiple solutions. Therefore, in the later, final design phases, there are generally fewer iterations, and fewer opportunities to regress to the earlier design phases. It is important that tools which operate during the conceptual and design development phases, allow the designer the creative freedom to thoroughly explore the problem design space without being prematurely directed toward a conclusion. It is possible that by suggesting a single 'optimal' solution to the designer in the early design phases, a design aid (computer assisted or traditional) can actually hinder the complete exploration of the problem by not promoting the consideration of sufficient options. The concept of the IGDT is to aid the designer specifically in the early phases of a project by allowing the exploration of a multiplicity of solutions.

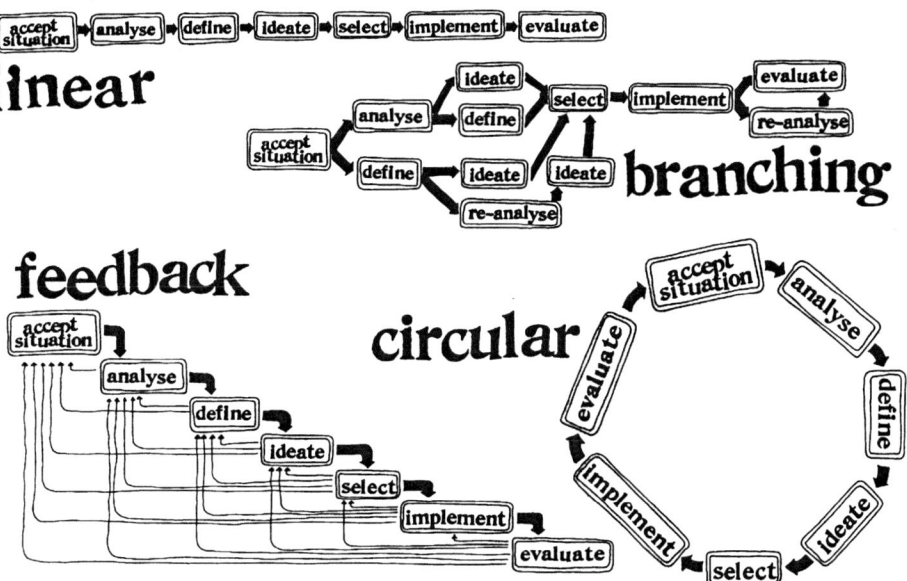

Figure 1.2. Four variations of design cycle patterns by Koberg and Bagnall (Koberg & Bagnall, 1972).

Although different disciplines do display variation in the description of the cycling process as is shown in Section 1.2.2.1., they share, nonetheless, a similar pattern. In their widely used design primer, *The Universal Traveler - a Soft-Systems guide to: creativity, problem-solving, and the process of reaching goals*, Don Koberg and Jim Bagnall propose seven basic steps to describe the cycle in the discipline of Architecture:

- Accept
- Analyse
- Define
- Ideate
- Decide
- Select
- Evaluate (Koberg and Bagnall, 1972)

Figure 1.2 shows that although the steps are progressive, the overall process remains cyclic with the possibility of regressive links at each step. As shown in the diagrams the ordering is not necessarily sequential, and varies from project to project. The variation in the four example flow diagrams in Figure 1.2, demonstrates the need for flexibility in an IGDT. In order to fit the work patterns of different designers, a useful design tool needs to be able to adapt to personal and problem specific patterns.

1.2 Study of Design

1.2.1 Theory and Practice of Design

According to Thomas Kuhn's definition, the field of design is not a mature science as it lacks a "coherent tradition of scientific research and practice, embodying law, theory, application, and instrumentation" (Kuhn, 1962). There exists no comprehensive theory that can be called upon to predict the outcome of a given design problem. For example, using the theories current in the science of astronomy, one can predict, within the limits of the theory, the position of a planet in the sky on a given day at a given time. But there are no such theories that can be used to predict the outcome of a design problem with such success. Kuhn describes through numerous historic examples the events which lead to the establishment of a new branch of science. For example the following paragraph cited from Kuhn's book, *The Structure of Scientific Revolutions*, shows how a field of endeavor can move from a desperate collection of independent conjectures to a unified science whose body of knowledge is directed by an accepted theory and laws.

> At various times all these schools made significant contributions to the body of concepts, phenomena, and techniques from which Newton drew the first nearly uniformly accepted paradigm for physical optics. Any definition of the scientist that excludes at least the more creative members of these various schools will exclude their modern successors as well. Those men were scientists. Yet anyone examining a survey of physical optics before Newton may well conclude that, though the field's practitioners were scientists, the net result of their activity was something less than science. Being able to take no common body of belief for granted, each writer on physical optics felt forced to build his field anew from its foundations. In doing so, his choice of supporting observation and experiment was relatively free, for there was no standard set of methods or of phenomena that every optical writer felt forced to employ and explain. Under these circumstances, the dialogue of the resulting books was often directed as much to the members of other schools as it was to nature. That pattern is not unfamiliar in a number of creative fields today, nor is it incompatible with significant discovery and invention. It is not, however, the pattern of development that physical optics acquired after Newton and that other natural sciences make familiar today.
> (Kuhn, 1962, <1996 ed., p. 13.>)

This Kuhnian view of development in the natural sciences is often overlaid on the design sciences, particularly in architecture and engineering. The result is a tendency to expect design to fit the mold of the natural sciences, i.e., to develop a unifying theory and be governed by application of its laws. Professor William Addis, historian and engineer in the Department of Construction Management at the University of Reading, has charted the development of this attitude in the field of engineering design, through numerous citations of engineers from the beginning of the 19th century to the present (Addis, 1990, pp. 3-13). Addis labels the natural science element of engineering as "theory" and the design element as "practice". This is certainly a popularly understood division of the field. Addis recognizes two camps in the field of engineering - those wishing to see theory as the dominant element, and those wishing to see design as dominant. The following quotation cited by Addis give a good representation of the position of the first camp.

> It is largely due to the efforts of ... the minority who devoted themselves to the theoretical aspects of the professional work ... in the past that the practical engineer has succeeded in gaining and , more important, maintaining the professional status which ... he values so highly. The theory of today should be the practice of tomorrow and unless theoretical knowledge is ever in advance of current practical requirements the

> survival of the engineer as a professional man is in danger ... The practitioners [must be] educated to understand and translate into reality the work of the scientist. (Pippard, 1956, p. 161)

Addis goes on to portray a second camp in the field of engineering, which he describes as follows:

> There is, however, a second, entirely different view of the role of theory in design. Its advocates consider theoretical calculations to be of secondary importance to a different type of knowledge. This other type of knowledge is a qualitative on, based upon an understanding of how materials and structures behave, rather than upon the abstract principles, laws or theories which are supposed to govern their behaviour. The names of Torroja, Nervi, Candela and Maillart are often associated with such views. (Addis, 1990, p. 10)

To support this view, Addis goes on to quote from the engineers he lists. Some of his citations are worth reproducing here to make the point clear.

> The calculations of stresses can only serve to check and to correct the sizes of the structural members as conceived and proposed by the intuition of the designer. (Torroja, 1967, p. 331)

> The most advanced chapters of theory of structures ... can only be used to check the stability of a structure. They can be used only to analyze numerically a structure already designed, not only in its general outline , but in all its dimensional relations. The formative stage of a design, during which its main characteristics are defined and its qualities and faults are determined once and for all (just as the characteristics of an organism are clearly defined in the embryo), cannot make use of structural theory and must resort to intuition and schematic simplifications. (Nervi, 1956, p. 17)

> It is admittedly a fairly widespread opinion that the dimensions [of a structure] should be unequivocally and finally determined by calculation. However in view of the impossibility of taking into account all possible contingencies, any calculation can be nothing but a guidance to the designer. (Maillart, cited in Straub, 1952, p. 240)

Addis comes to the conclusion that practical design depends on knowledge other than that embodied by scientific theory and laws. He lists some possible examples of design knowledge as:
- rules of thumb
- the numerous empirical data and rules associated with Codes of Practice
- the properties of particular materials
- factors of safety
- intuitive knowledge of structural behaviour
- experience
- engineering judgment   (Addis, 1990, p. 11)

Nobel Laureate and Professor of Computer Science and Psychology at Carnegie Mellon University, Herbert Simon, makes the distinction clear between theory and practice in his book *The Sciences of the Artificial*. Whereas what Addis refers to as engineering theory is derived from the natural sciences, the practice of design in engineering and architecture belongs to the sciences of the artificial.

> Historically and traditionally, it has been the task of the science disciplines to teach about natural things: how they are and how they work. It has been the task of engineering schools to teach about artificial things: how to make artifacts that have desired properties and how to design. (Simon, 1969, <1996 ed., p. 111>)

> The natural sciences are concerned with how things are. ... Design, on the other hand, is concerned with how things ought to be, with devising artifacts to attain goals. (Simon, 1969, <1996 ed., p. 114>)

Simon comments further on the rift between theory and practice, elucidated above by engineers of both camps. He describes the effect this factious spirit has had on colleges of engineering during the past century.

> In view of the key role of design in professional activity, it is ironic that in this century the natural sciences almost drove the sciences of the artificial from the professional school curricula ... Engineering schools gradually became schools of physics and mathematics ... The use of adjectives like "applied" concealed, but did not change, the fact. It simply meant that in the professional schools those topics were selected from mathematics and the natural sciences for emphasis which were thought to be most nearly relevant to professional practice. It did not mean that design continued to be taught, as distinguished from analysis. ...
> The stronger universities were more deeply affected than the weaker, and the graduate programs more than the undergraduate. During that time few doctoral dissertations in first-rate professional schools dealt with genuine design problems, as distinguished from problems in solid state physics or stochastic processes. ...
> As professional schools, including the independent engineering schools, were more and more absorbed into the general culture of the university, they hankered after academic respectability. In terms of the prevailing norms, academic respectability calls for subject matter that is intellectually tough, analytic, formalizable, and teachable. In the past much, if not most, of what we knew about design and about the artificial sciences was intellectually soft, intuitive, informal, and cookbooky. ...
> The damage to professional competence caused by the loss of design from professional curricula gradually gained recognition ... Some schools did not think it a problem (and a few still do not), because they regarded schools of applied science as a superior alternative to the trade schools of the past. If that were the choice, we could agree. But neither alternative is satisfactory.
> The professional schools can reassume their professional responsibilities just to the degree that they discover and teach a science of design, a body of intellectually tough, analytic, partly formalizable, partly empirical, teachable doctrine about the design process. (Simon, 1969, <1996 ed., pp. 111-113>)

Simon goes on to describe how the introduction of computers into the design process has forced design theoreticians to formally and explicitly describe aspects of the design process. This in turn has given design a more academic acceptable image.

The following sections outline some of the directions the study of design and design theory have recently taken. As opposed to the study of natural science, where one can expect to see a unification of thought supporting a dominant theory, no such Kuhnian paradigm should be expected in the study of design. As a science of the artificial, design is a very different animal. Design is not constrained by attempts to describe the single reality "what is", but left totally open to describe the "what ought to be". It brings with it the frustration of no single touchstone of verification, but offers instead the opportunity to express that which lies beyond the natural.

This lack of a single touchstone also complicates the study of design in that no single set of goals guide the formation of procedures. Where some practitioners can offer more compete theories of their work, others describe at best proven techniques. The remainder of this section is divided into two sections. Section 1.2.2., Design Models, attempts to delineate some of the more prominent theories current in design, and Section 1.2.3., Design Mechanisms, catches other aspects of design which are recognized as truisms, independent of a specific theory.

1.2.2 Design Models

If one cannot expect a unified theory of design, it is certainly possible to have multiple working models that describe design in some useful ways. Case studies are commonly studied as well in an attempt to recognize what it is that brings success to a working design methodology. Since design describes "what ought to be" there are, in fact, as many models for design as there are views of what ought to be. For example, design has been modeled as language (Semiotics), as a logic system (Archer, 1970), as a problem solving process (Newell, Simon, 1972), by procedural methods (Asimow, 1962), by reference to typologies (Moneo, 1978), or by simulation and optimization techniques. Each of these models is able to offer a particular view of design, and their study can be enriching. In the sections that follow, a few of the design models more significant to the development of the IGDT are outlined. They include:

- Design as Process
- Design as Simulation
- Design as Optimization
- Design as State Space Exploration

1.2.2.1 Design as Process

Much effort has been applied to develop a method which describes design as a prescriptive process. By modeling design as a series of steps the designer is provided with a method which not only gives structure to the activity, but helps to sharpen the general understanding of what design is. In his 1967 book, *Creativity: The Magical Synthesis*, Silvano Arieti discusses eight models of creative design published between 1908 and 1964. In reviewing several models it is interesting to observe the commonality present, as well as the aspects stressed by different authors. Paul E. Plsek, founder of Directed Creativity, updates Arieti's list of procedures in his 1997 book *Creativity, Innovation, and Quality*. Plsek charts the development of two schools of thought in creative methodology. The first group holds that creative design is primarily a subconscious act, and therefore does not lend itself to description by prescriptive steps. The second group agrees that there may be a subconscious component in design, what might be called inspiration, but hold that the process as a whole is certainly a conscious, rational and describable act.

> It is important to note that some experts dismiss the notion that creativity can be described as a sequence of steps in a model. For example, Vinacke (1953) is adamant that creative thinking in the arts does not follow a model. In a similar vein, Gestalt philosophers like Wertheimer (1945) assert that the process of creative thinking is a integrated line of thought that does not lend itself to the segmentation implied by steps of a model. But while such views are strongly held, they are in the minority.

> Business people, who have used models for quality improvement, strategic planning, reengineering, and so on, are well-positioned to deal with this apparent controversy. We understand, by experience, that while models are helpful in guiding our efforts, they are not to be used too rigidly. We understand that models are not rote prescriptions. We may deviate substantially from a model in a given situation, but this does not render the model useless. We also understand the concept of flow and realize that one should not be to dogmatic about when one step of the model ends and the next begins. Models are useful, but only a fool follows them blindly. (Plsek, 1997)

One of the earliest models discussed by both Plsek and Arieti is that of Graham Wallas (1926). Wallas, whose is often referenced in creative thinking programs, used a four step design model.

- Preparation definition of issue, observation and study
- Incubation laying the issue aside for a time
- Illumination the moment when a new idea finally emerges
- Verification checking it out

Wallas would seem to combine the two schools described by Plsek, the intuitive and the analytic. Although incubation and illumination seem to be primarily intuitive, and guided by the subconscious, Wallas begins and ends his model with decidedly analytic steps, observation, study and verification.

Two proponents of the intuitive school are Dean Keith Simonton, Professor of Psychology at the University of California, Davis, and author of *Origins of Genius: Darwinian Perspectives on Creativity* (1999); and Donald T. Campbell, former Professor Emeritus of Sociology, Anthropology, Psychology and Education at Lehigh University, originator of the term "Blind Variation and Selective Retention" (BVSR), likewise in reference to Darwinian evolution. In the field of creative design, both men advanced the "chance configuration theory", which has its origins in the writings of the psychologist William James in the 1880's. The chance configuration theory holds that creativity in design comes about to a large extent though uncontrollable elements guided by random chance. The configuration of experiences or knowledge that any person might bring to bear on a design problem is largely guided by chance, and successful solutions are selectively retained. A selected solution is refined by the designer, and undergoes a process of adaptation in the manner of Darwinian evolution. The final solution survives to be implemented by the designer. Although the initial step is guided by chance and intuition, retention and development of a chance solution are certainly analytic in nature. A further discussion of the roll of chance in design is given in Section 1.2.3.1.

An even greater emphasis is placed on the roll of subconscious and chance processes in the design model proposed by F. Barron (1988).

- Conception in a prepared mind
- Gestation   time, intricately coordinated
- Paturation   suffering to be born, emergence to light
- Bringing up the baby   further period of development

Called the "psychic creation model", Barron's four phase plan leaves the first three phases beyond direct access in the designer's subconscious.

But in contrast to Barron-type "psychic" models, several models cited by Plsek give a more balanced view of the rolls of intuition and analysis (Plsek, 1997). Rossman, for example, after compiling surveys of 710 inventors, refined Wallas's four step model as follows:

- Observation of a need or difficulty
- Analysis of the need
- A survey of all available information
- A formulation of all objective solutions
- A critical analysis of these solutions for their advantages and disadvantages
- The birth of the new idea - the invention
- Experimentation to test out the most promising solution, and the selection and perfection of the final embodiment

Although Rossman also contains an element of mystery in "the birth of the new idea", the preceding and following steps give a more analytical balance missing in Barron's model.

The originator of the well known technique of "brainstorming", Dr. Alex F. Osborn, also published a seven step model for creative thinking in his book, *Applied Imagination* (1963).

- Orientation pointing up the problem
- Preparation gathering pertinent data
- Analysis breaking down the relevant material
- Ideation piling up alternatives by way of ideas
- Incubation letting up, to invite illumination
- Synthesis putting the pieces together
- Evaluation judging the resulting ideas

Osborn's model is even more pragmatic, but includes the need for illumination, which can be supported by ideation, but in the end occurs more intuitively during incubation. In 1954, Dr. Osborn founded the Creative Education Foundation at the University of Buffalo, which hosted the first Creative Problem Solving Institute (CPSI) that same year. CPSI has continued to provide workshops each year, teaching the CPS method to tens of thousands of professionals and educators in the United States. Dr. Sidney J. Parnes, the first director of CPSI, along with other CPSI faculty including Dr. Donald Trefflinger and Dr. Scott Isaksen have published extensively on the CPS method (Parnes, 1992; Isaksen & Trefflinger, 1985). The CPS method incorporates the following steps:

- Objective finding
- Fact finding
- Problem finding
- Idea finding
- Solution finding
- Acceptance finding

Several other educators in the field of creative design methodology in the United States have philosophic ties with CPS, and have proposed variations of the CPS method. Dr. Win Wegner, director of Project Renaissance, combines the first three steps defined by CPS in his "Gravel Gulch" four step method (Wegner, 1981)

- The mess
- Idea-finding
- Solution-finding
- Action-planning

Another popular model is the seven step method by Don Koberg and Jim Bagnall (1972), discussed in Section 1.1.2.2.

- Accept the situation
- Analyze
- Define
- Ideate
- Select
- Implement
- Evaluate

Figure 1.2, which shows possible flow charts of the method, make it clear that the process is not seen as strictly linear, but can contain multiple layers of recursion. In the tradition of CPS, the model of Koberg and Bagnall brackets the intuitive phase of ideation by the analytic phases at the beginning and the end of the process.

In the area of architectural engineering, Dr. Morris Asimow is credited with advancing the basic design cycle:

- analysis
- synthesis
- evaluation

In his book, *Introduction to Design* (Asimow, 1962) Dr. Asimow describes how the three phases are interconnected.

> A philosophy of engineering design comprises three parts, namely, a set of consistent principles and their logical derivatives, an operational discipline which leads to action, and finally a critical feedback apparatus which measures the advantages, detects the shortcomings, and illuminates the directions of improvement. (Asimow, 1962, pp. 4-5)

> The design process resembles the general process of problem solving in the main features, but it uses sharper, and for the most part, more analytical tools, which have been especially shaped and sharpened for the problems of engineering design. It carries the process through

> *analysis, synthesis,* and *evaluation and decision,* and extends it into the
> realms of *optimization, revision,* and *implementation.*
> (Asimow, 1962, p. 44)

Paul E. Plsek (1997) adds to Asimow's basic triad the CPS element of acceptance, which he labels "living with it". Plsek suggests that this critical step forms the link that couples the end with the start of the process. Figure 1.3 shows the Directed Creativity Cycle by Plsek.

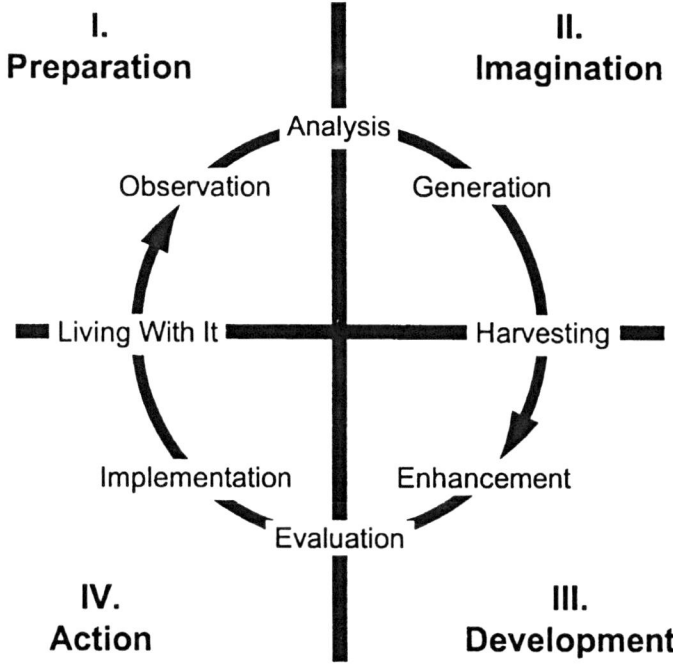

Figure 1.3. The Directed Creativity Cycle developed by Paul E. Plsek.

Plsek gives the following short description of the Directed Creativity Cycle.

> We **live everyday** in the same world as everyone else, but creative thinking begins with careful **observation** of that world coupled with thoughtful **analysis** of how things work and fail. These mental processes create a store, we **generate** novel ideas to meet specific needs by actively searching for associations among concepts. There are many specific techniques that we can use to make these associations; for example, analogies, branching out from a given concept, using a random word, classic brainstorming, and so on. The choice of technique is not so important; making the effort to actively search for associations is what is key.
>
> Seeking the balance between satisficing and premature judgment, we **harvest** and further **enhance** our ideas before we subject them to a final, practical **evaluation**. But, it is not enough just to have creative thoughts; ideas have no value until we put in the work to **implement** them. Every new idea that is put into practice changes the **world we**

live in, which re-starts the cycle of observation and analysis.
(Plsek, 1997)

Finally it is worth considering the procedure forwarded by Herbert Simon (1973) in discussing "ill structured problems" as compared to "well structured problems". The distinction is that well structured problems can be solved by general problem solving procedures in a straight forward manner (i.e., analysis), whereas ill structured problems cannot be solved so directly. Simon builds the argument that *all design problems* are ill structured because at the outset of the process the problem space is not fully specified, in fact some parameters of the problem may only "occur" to the designer after considerable work and search. In specific regard to architectural design (with the proviso that "creative" design is intended), Simon writes:

> The design task (with this proviso) is ill structured in a number of respects. There is initially no definite criterion to test a proposed solution, much less a mechanizeable process to apply the criterion. The problem space is not defined in any meaningful way, for a definition would have to encompass all kinds of structures the architect might at some point consider (e.g, a geodesic dome, a truss roof, arches ...), all considerable materials (wood, metal, plexiglass, ice ...), all design processes and organizations of design processes (start with floor plans, start with list of functional needs, start with façade, ...).

Simon outlines a procedure for solving these ill structured problems (Simon, 1973). Simon also recognizes that the approach is cyclic and alternates between solving component parts of the problem and attaining and assimilating new information about the problem. Unlike analysis, it is not necessary for the process to be completely defined by constraints. In fact, Simon observes, "The more distinguished the architect, the less expectation that the client should provide the constraints." (Simon, 1973)

As will be seen in later sections, there are many aspects of Genetic Algorithms which parallel many of these procedures. The design procedures discussed above are generally all seen as cyclic, with cycles occurring within each step. This is the same structure which Simon terms "hierachtic" (Simon, 1969, <1996 ed., p.184>). The cyclic nature of Genetic Algorithms can be seen as very similar (v. Bülow, 2007). It is also seen that the cyclic steps converge on a solution. This is also the expectation with Genetic Algorithms. Design procedures usually allow for the possibility of return to earlier steps with additional information gained from later steps. This is an important feature of an IGDT discussed in Section 1.3.3. Flow charts similar to the ones described by Koberg and Bagnall in Figure 1.2, could very well apply to the flow of an IGDT.

Finally, the positioning of the IGDT within the design models discussed above is important to note. The IGDT is intended to aid the designer in the intuitive, ideation phases of design. In Plsek's model, this is "Generation". In the Koberg and Bagnall model it comes as "Ideate". Although supported before and after with some form of analysis, the aid offered to the designer is in the area of intuition. In this area the IGDT is fairly unique as a design tool.

1.2.2.2 Design as Simulation

In simulation the aspects of the design which are to be considered are in some way modeled. This model may be a physical model, a mathematical model or even a rather abstract thought model. In any case our limited understanding of systems usually prevents us from producing a simulation which accurately copies the original in every detail. Simulation is not replication. But it is assumed that if enough of the more important parameters are incorporated in the model, the simulation will react to testing in a way that exposes more information about the system than could easily be observed otherwise.

Physical models have a long history of providing richly rewarding simulations. They are used extensively in architecture and engineering to model aspects of form, space and structure. Known laws of similitude provide a means of quantitative translation of physical behavior under various internal or external loadings.

Simulation is probably the most commonly used approach to solving a design problem. In simulation, a trial design is described, and its performance is calculated or in some way predicted. Actually, simulation may be applied to either a physical model which can be tested and observed, or to a mathematical model used in calculations. Depending on how accurately the design is modeled, simulations can give very good results. The final product of a design is, in a sense, the ultimate simulation at 1:1, from which much information may be learned for future designs.

Simulation has, however, two significant draw backs. First, it is not always clear, once the performance is found lacking, what parameters should be altered, and how they should be altered, to affect the performance. For instance, once an element is modeled or produced it may be known very accurately how much it weighs, but that may not provide much direction in actually making it lighter. The second draw back can be in the efficiency in terms of time. With limited guidance the simulation method may result in much time consuming trial and error. If physical models are employed, this will greatly limit the number of iterations of design trials which can reasonably be executed.

1.2.2.3 Design as Optimization

Because no designer deliberately chooses a solution that poorly fits the design criteria, all design can be seen as optimization in some form or other. For this reason optimization methods have received much attention in many different fields that employ design. Mathematical methods of optimization include many techniques, the best known being Linear Programming. Common to all such methods is the necessity to be able to describe the design parameters as mathematical variables. The domain from which possible variable values are taken is constrained by initial design decisions. These initial design decisions are themselves variable, the decision variables, and are usually chosen by the designer. In his paper, "The Structure of Ill Structured Problems" (Simon, 1973), Herbert Simon points out that any problem in which the initial design decisions change due to the introduction of "new resources that 'occur' to (the designer) in the course of his solution efforts" is an ill structured problem. As a result, design problems are usually converted to 'well structured' analysis problems for optimization, by fixing the decision variables.

Performance variables point to a chosen design solution which can be found in the domain described by the decision variables. The desired design is described through objectives, which for the optimization procedure must be defined as mathematical functions. Applying a set of performance variables to the objective functions yields the design solution. The range defined by all possible solutions which might be found by applying any possible combination of performance variable values to the objective functions is the solution space. The optimization problem becomes a process of finding the combination of performance variable values, that when applied to the objective functions, points to some best (or worst) design solution.

Although developments in optimization techniques can be traced as far back as Galileo de Galilei's investigations of the cantilever beam in 1638 (Seireg & Rodriguez, 1997, p.1), the development of the linear programming method by Dantzig in 1948, and variations to the simplex method developed during the 1950's, issued a renewed interest in optimization techniques (Venkayya, 1993, p.2). Schmit's work with the structural synthesis technique in the late 1950's (Schmit, 1960), introduced many structural engineers to the field of optimization. In the 40 years since then, both nonlinear extensions to linear programming methods as well as the development of nonlinear programming algorithms have advanced the application of optimization techniques. The historic development of optimization techniques has been well documented by the authors cited above as well as numerous others. Still, such methods are not commonly used in architectural engineering. The formulation of problems is not simple and the results are not always immediately useful to designers.

The formulation of most optimization methods can be described by the following terms:

- decision variables
- performance variables
- objective functions

Decision variables set the context for the design. They are the initial parameters decided upon by the designer which qualify the design space. That is, by defining the decision variables, the designer sets the limits or constraints which specify a family of possible solutions. This family of solutions can be thought of as being contained in some (n dimensional) space. This space has as many dimensions as there are decision variables. In architectural engineering problems the number of variables can be quite a few. Each of these multidimensional spaces can possibly contain an infinite number of possibilities. Searching through such spaces with simple, unguided trial-and-error is not likely to find the best solution.

Performance variables define particular solutions within a design space. Together they form a code which allows each solution to be accessed. Conceptually, they can be seen as providing the address of each individual in the design space. Each combination of performance variables points to a specific solution in a similar way that addresses composed of country, state, city, street, and number, point to a specific individual on the planet Earth.

Objective functions define the particular solution of interest in the design space. The solution that best fits the objective criteria is the 'optimal' solution which is sought. In the optimization process, the goal is to find the combination of performance variables which point to a solution in the design space which displays the optimal combination of objective criteria. The performance variables provide the means to point to a solution, whereas the objective functions describe the solution that is desired.

Of course, a designer usually considers more than one, single criterion. Many criteria may overlap or even conflict. In that sense there is often no solution which can optimize *all* design criteria. So one roll of the designer becomes the arbitration of which criteria to consider at all, and which of those impart a greater importance. But a major problem with the various mathematical methods of optimization when applied to architecture lies in the inability to formulate the objectives in quantifiable, mathematical terms. In

reviewing optimization methods in relation to knowledge based systems Coyne, et al. write:
> The main problem with most of the models used in design methods, especially in optimization, is that they can deal only with quantifiable parameters stated in mathematical terms. (Coyne, et al., 1990, p.28)

As a result, only objectives that can be written as functions are considered, and then those that are considered are usually simplified in writing the objective function. Thus, the 'optimum' that is found, is not actually the optimum for the real problem, but only for the limited and simplified problem. Nonetheless, an optimum found in this way can give satisfactory results if enough of the primary objectives are reasonably defined.

Simon suggests, that rather than trying to find 'optimal' solutions to the simplified problem, one might better seek satisfactory solutions to the real problem (Simon, 1969, <1996 ed., p.26>). This is the approach taken with heuristic techniques used in Artificial Intelligence (AI) methods. However, although AI methods operate on a more realistic model of the problem, the heuristic objectives still have to be stated and coded into the programs. An IGDT as described in this book offers the designer a great deal more flexibility with respect to objective functions, in that they do *not* have to be quantifiable statements coded into the program. Like Simon's description of AI, an IGDT seeks "satisficing" solutions. As described in Section 1.3.3.2., an IGDT is steered by the designer interactively, using whatever quantitative or qualitative criteria available.

1.2.2.4 Design as State Space Exploration

Although, as noted above, design is goal oriented, and thus often modeled as problem solving, it differs from problem solving in that the goals are not fixed at the outset. In fact, determining the goals is part of the process. Section 1.2.3.2. discusses how goals can change as the level of knowledge brought to the process increases. In taking this into account, Gero (1994, p. 315) and other authors (Newell & Simon, 1972, p. 76) somewhat shift the model of design as problem solving to that of design as a decision making activity. Decision making implies making choices, and choices are framed by parameters that can be described as variables. In ordering these variables, Gero (1994, p. 316) represents design as being comprised by three state spaces:

- Function   (definition of objects purpose - teleology)
- Behavior   (performance space)
- Structure (decision space)

Search methods of different types can be used to find values of variables in a state space, but Gero suggests, that more critical to design is the determination of the state space within which to search. This he terms *exploration* of the state space. In the process of exploring, learning takes place as new knowledge is gained or old knowledge restructured.

Structure is described by Rosenman & Gero (1997) as the "what is". It represents the physical object itself as described by material, topology, geometry and physical characteristics. Structure contains the variables for the parameters necessary to replicate the object. These are the parameters that a designer varies to describe or "decide" a particular design solution.

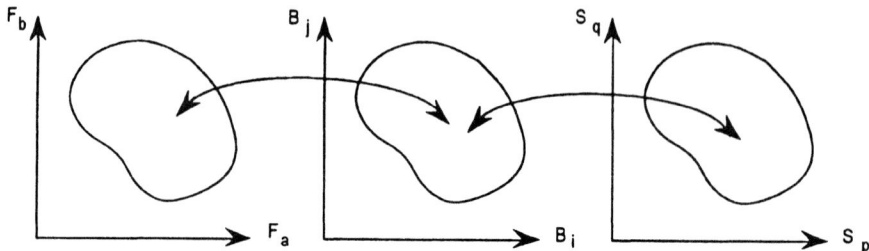

Figure 1.4. The three design subspaces of function = F, behavior = B, and structure = S. (Gero, 1994, p. 317).

Behavior is the "how does". It is the description of how the design behaves in a particular environment. Deformation, resilience, ductility, creep, are typical behaviors for architectural engineering designs. Behavior can usually be mapped to Structure. This allows specific structures to be sought that link-to or produce certain behaviors. Figure 1.5 shows this behavior mapping used to search for a structure.

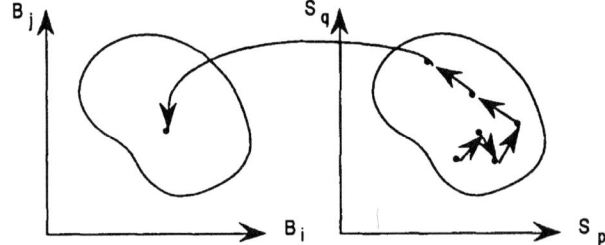

Figure 1.5. Search for a specific structure that maps to a certain behavior, (Gero, 1994, p. 318).

Function is the "what does". It is distinct from *purpose* (the "why does"), and can be seen as the result of behavior. In this sense it is closely linked to behavior. But it is often difficult, if not impossible, to link function and structure directly (see no-function-in-structure principle (de Kleer & Brown, 1984)). Despite Louis Sullivan's much popularized dictum of "form follows function", (in this case "form" being "structure"), the link between form and function usually depends on context.

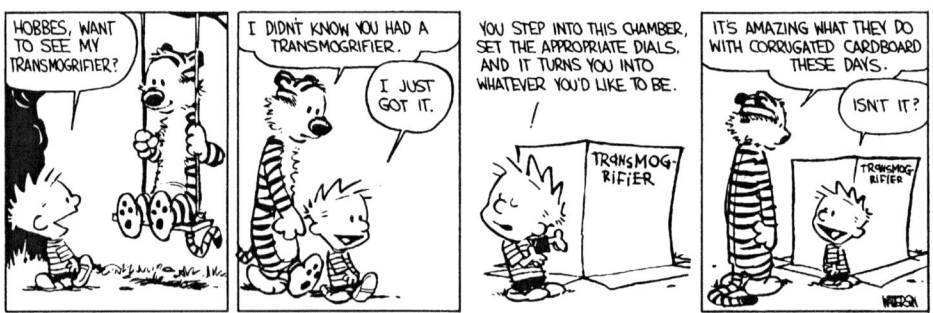

Figure 1.6. An example of one "function" of a cardboard box, (Watterson, 1988, p. 229).

Regarding these three divisions of variables as state spaces, it can be seen that structure can be mapped to behavior, and behavior can be mapped to function, but it is usually not possible to map structure directly to function or vise-versa (Gero, 1994, p. 316). For example, a corrugated cardboard box has a certain structure. This structure describes a specific, predictable behavior - e.g., burst strength of 10 pounds. The behavior can be seen as allowing a function - e.g., packaging 10, 1 pound books. But given the structure alone, there is no unique mapping to function. In the hands of a five-year-old, the same structure could be split on the seam and used as a means of transportation down a snowy bank, or opened on both ends to provide a tunnel, or inverted with lettering to function as a transmogrifier (Watterson, 1988, p. 229). Attempts to directly map structure to function are at best prone to failure and at worst would severely inhibit innovation that is so critical to good design.

1.2.3 Design Mechanisms

There are several mechanisms that are commonly recognized in one form or another as being able to help, or if neglected, hinder, the creative thinking necessary for conceptual design. This section discusses the following as representative of the more import mechanisms.

- Chance in Design
- Knowledge in Design
- Fixation in Design
- Search and Exploration in Design
- Emergence in Design
- Lateral Thinking in Design

These mechanisms are for the most part known to persons practicing design, but, nonetheless, usually not accounted for in the development of computer design tools. Mechanisms like these are no doubt described in every introductory design course. However, when one looks for evidence of there application in current computer based design aids, they are noticeably lacking. This omission is certainly the source of apprehension many designers have toward using computer programs to aid conceptual design. For a design tool to be successful in promoting creative thinking, careful consideration needs to be given to how the program will support such mechanisms. This section describes some of the more important of these mechanisms and gives indications of how they are incorporated in an IGDT.

1.2.3.1 Chance in Design

Every designer can cite anecdotes of how seeming happenstance played a key role in the development of a design concept. As a design tool, chance is often disregarded as being an uncontrollable element. In his book *Chase, Chance and Creativity*, James Austin, Professor Emeritus at the University of Colorado Medical Center and former Chair of the Department of Neurology, defines four types of chance, and explains how they relate to creative problem solving (Austin, 1978). Through numerous examples he shows how successful scientists and designers tend to increase the likelihood of chance occurrences or suggestions which have a positive influence on their work.

Various Aspects and Kinds of Good Luck

Term Used to Describe the Quality Involved	Good Luck is the Result of ...	Classification of Luck	Elements Involved	Personality Traits You Need
	An Accident	Chance I	"Blind" luck. Chance happens, and nothing about it is directly attributable to you, the recipient	None
SERENDIPITY	General Exploratory Behavior	Chance II	The Kettering Principle. Chance favors those in motion. Events are brought together to form "happy accidents" when you diffusely apply your energies in motions that are typically nonspecific.	Curiosity about many things, persistence, willingness to experiment and to explore.
	Sagacity	Chance III	The Pasture Principle. Chance favors the prepared mind. Some special receptivity born from past experience permits you to discern a new fact or to perceive ideas in a new relationship.	A background of knowledge, based on your abilities to observe, remember, and quickly form significant new associations.
ALTAMIRAGE	Personalized Action	Chance IV	The Disraeli Principle. Chance favors the individualized action. Fortuitous events occur when you behave in ways that are highly distinctive of you as a person.	Distinctive hobbies, personal life styles, and activities peculiar to you as an individual, especially when they operate in domains seemingly far removed from the area of the discovery.

Table 1.1. Austin's Classification of the Four Types of Chance (Austin, 1978, p.78)

Chance I Austin defines as "blind luck" (Austin, 1978, p. 73). Like winning a lottery, this is the type of chance that cannot be predicted or expected in any way, and has really nothing to do with the recipient. Although a possibility, it is statistically so unlikely that a useful solution would be found to any complex problem in this way, that it would not be of any real use in a design tool. A series of randomly generated solutions might at least be seen as stimulating, but lacking the control given by a GA in allowing directed development, such a tool would be of very limited value.

Chance II occurs in situations where by dint of perseverance and diligent, directed activity, the likelihood of a "happy accident" is greatly increased. As an example Austin offers the following words from the engineer and inventor, Charles Kettering (1876 - 1958):

> Keep on going and the chances are you will stumble on something, perhaps when you are least expecting it. I have never heard of anyone stumbling on something sitting down. (Austin, 1978, p. 72)

This is the type of chance discovery that one seeks though exploration of different design spaces, and it is this type of discovery that an IGDT supports as an exploration tool. By the shear number of solutions that can be perused using an IGDT, the chances are increased that "you will stumble on something" in the process. The difference between this type of luck and "blind luck" (Chance I) is that you find it rather than it finding you.

Chance III Austin characterizes with a quote from Louis Pasteur: "Dans les champs de l'observation, le hazard ne favorise que les esprits préparés" (Chance favors only the prepared mind) (Austin, 1978, p. 74). Although the same solution may be presented to different individuals, it is a specific designer's own skill and background that allows for the recognition of the worth of that solution. This is also a key factor in the IGDT concept. Unlike many analysis tools that seek to distill an optimum solution given only a set of initial parameters, an IGDT requires the continued interaction with a human designer to provide "les esprits préparés" (the prepared mind) that can recognize and select better solutions.

Austin characterizes these first three forms of chance as having the quality of serendipity - a term originally coined by Horace Walpole indicating the kind of luck one has for unexpectedly encountering a solution either through sagacity or accident (Austin, 1978, p. 71). To characterize Chance IV Austin coins his own term, "Altamirage", which he derives from the lucky discovery by Don Marcelino of the cave paintings in Altamira dating to the Magdalenian era of the Old Stone Age (between 15,000 BC and 12,000 BC).

Chance IV is dependent on the specific qualities (hobbies, diverse interests, life style) of the person. By "Altamirage" Austin indicates that this type of luck only happens through the chance coming together of the specific qualities of an individual (Austin, 1978, p. 77). In the case of Don Marcelino, the facts that he had interests in history, local geology, had seen an archeology exhibit at the International Exhibition of 1878, liked the out-of-doors, and had a small and inquisitive daughter, all combined to make for the chance discovery of the cave paintings. Austin summarizes these four types of chance as shown in Table 1.1.

1.2.3.2 Knowledge in Design

W. J. Fabrycky has shown that there is a relationship between knowledge about a design and freedom to alter that design (Fabrycky, 1991). As work on a problem progresses toward a solution, the degree of freedom to explore radically different options lessens.

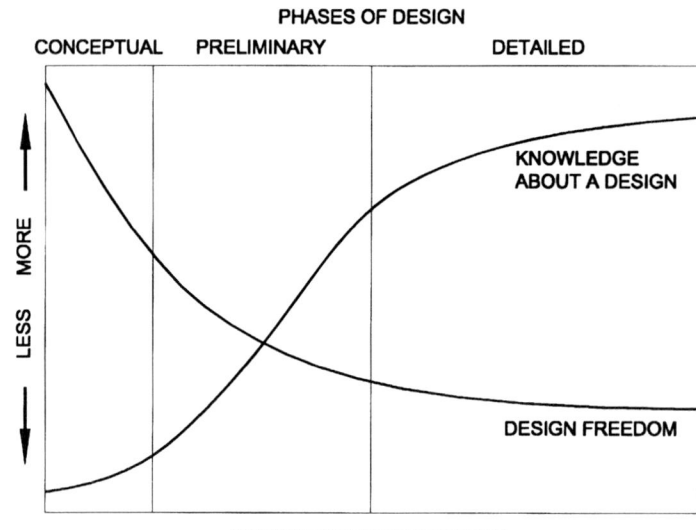

Figure 1.7. Typical Design Freedom and Knowledge vs. Time from W. J. Fabrycky.

Each decision made by the designer brings with it more definition to the problem, and thereby more knowledge (quantitative data) about the nature of the problem. But at the same time as each decision acts to narrow the scope of the problem, the degree of freedom to explore alternate directions and possibilities decreases. W. J. Fabrycky graphs knowledge about a design versus design freedom as shown in Figure 1.7 (Fabrycky, 1991). From this graph one can see that design freedom is of major importance in the conceptual design phase and early preliminary design phase, whereas knowledge about the design gains importance, and is primarily dominant in the last detailed design phase.

De Bono offers a similar graph relating amount of information present to creativity. De Bono sees a certain amount of information as useful and necessary to stimulate creative thought. But a point can be reached where the designer can be bogged down by too much information.

> Data does not by itself generate ideas. It usually does not even suggest them. ...It is useless to believe that creative effort can be replaced by a careful accumulation of data. If ideas are needed, more data is no substitute. (de Bono, 1971, p. 183)

Because traditionally, computers have been applied to quantitative problems, design tools which utilize computers have focused on the later detailed design phase where quantitative knowledge is primarily dominant.

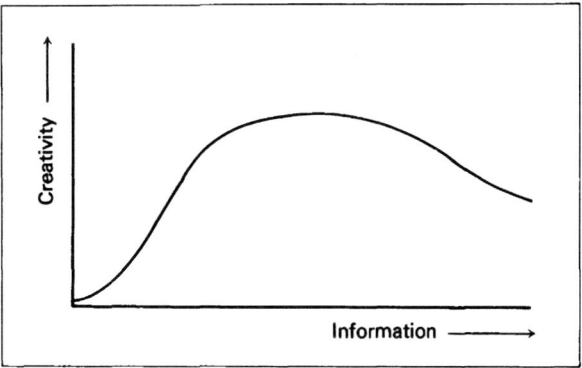

Figure 1.8. De Bono's Graph Relating Creativity and Information (de Bono, 1971, p. 184).

Efforts to utilize computers in earlier phases has often produced counter productive results by offering quantitative knowledge too early in the process. By offering the designer a quantitatively detailed solution in early design phases, there is the temptation for the designer to accept the computer generated solution (or at least that direction of consideration), without further exploration of design possibilities. Figure 1.9 shows the affect of a premature infusion of quantitative knowledge from a traditional computer design tool.

The immediate gain in knowledge results in the sudden jump in the graph during the early design phase. At the same time, with the acceptance of this knowledge, design freedom drops. Unfortunately, since knowledge is normally gained through the free exploration of the design space, with this freedom stifled, there is no base for a further increase in knowledge. As a result, the final design, although more quickly found, will likely not reach the level of success that would be expected *without* the computer input. In practice, of course, good designers know not to be lured into accepting the first solution they find (even if it is completely detailed by a computer). Good designers either use computer tools in the later design phases or have the discipline to ensure that the design space is adequately explored.

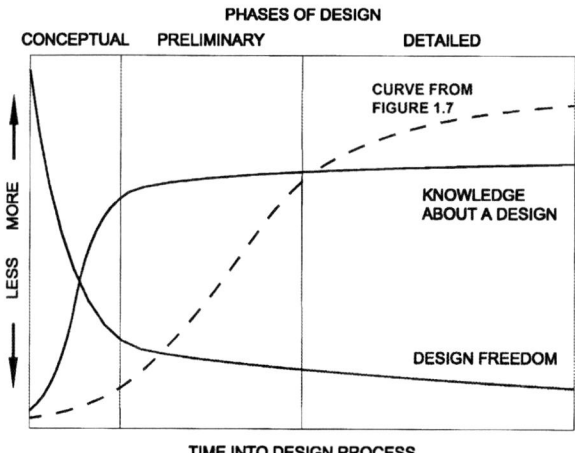

Figure 1.9. Design Freedom and Knowledge vs. Time with Premature Knowledge Infusion.

1.2.3.3 Fixation in Design

The phenomena of design fixation is well known, particularly to design educators, albeit usually under different names. Beginning design students are perpetually warned about the dangers of becoming overly enamored or stuck on an idea, to the exclusion of investigating or even being able to recognize other possibly better solutions. Osborn's brainstorming techniques (Osborn, 1963), Gordon's Synectics (Gordon, 1961), Adams' blockbusting (Adams, 1974), de Bono's lateral thinking (de Bono, 1971), as well as a host of other techniques and methods, all aim at helping the designer avoid becoming fixated on a single solution. Colbron, et al. (1993) and Purcell & Gero (1996) have pointed out how difficult it is for people to see new functions for common objects or new solutions to well known problems. De Bono presents this phenomena in terms of the patterning system of the human brain. In discussing patterning systems, de Bono uses the analogy of dripping a hot liquid (ideas or knowledge) onto a gelatin surface (the brain). Before the liquid cools, it will dissolve some of the gelatin and after it is poured away, a slight depression remains in the gelatin. The next time a drop of liquid falls on the gelatin surface, even though it may not fall at exactly the same spot, it will tend to flow toward the depression already present, thus enlarging it slightly. Eventually, erosion like rivulets are formed and new droplets that fall tend to roll into these rivulets rather than forming new depressions. Fixation effects ones thinking like these rivulets that steer the new droplets into pre-established patterns. Once the rivulets are present, it is very difficult for a new droplet not to run down an established rivulet, even though another course might be in some way better (shorter or whatever). The techniques mentioned above all try to overcome this phenomena in one way or another.

Being presented with a computer derived 'optimal' solution, can form a deep rut in ones cerebral gelatin, which tends to hinder the designer from further exploration of the design space. In addition, the over detailing of early design solutions, as commonly presented by computer based design aids, tends to have an effect similar to the excessive infusion of knowledge too early in the design process.

An IGDT responds to this problem by not offering *single*, optimized solutions, but rather *families* of good solutions. In addition to the advantage of robustness offered by a "satisficing solution model", as described in Section 1.2.2.3., the multiple solutions offered by an IGDT reduce the danger of design fixation. When one can view several solutions simultaneously, there is less danger of falling too deeply into one single path. In this way fixation is minimized and exploration is encouraged.

1.2.3.4 Search and Exploration in Design

John S. Gero describes exploration in design as follows:

> Exploration in design can be characterized as a process which creates new design state spaces or modifies existing design state spaces. New state spaces are rarely created de novo in design, rather existing design state spaces are modified. The result of exploring a design state space is an altered state space. (Gero, 1994, p. 318)

Search, on the other hand, refers to the computational process usually associated with optimization. It requires that the state spaces of behavior and structure be well defined. This would require that the variables which define these states be known as well as the functions that map one to the other. Specific values of structure variables are sought, which map to a certain set of behavior variables. Hale (1996) points out that since behavior variables may change during the process of conceptual design, some optimal solutions may become invalid. Figure 1.10 shows Hales graphic depiction of this situation. Hale proposes the use of solution "mesas", as shown in Figure 1.10, which provide a ranged set of solutions rather than single optimal solutions. The "satisficing solution model", originally proposed by Simon (1969) is described by Hale as capable of providing a more robust solution, but remains primarily a search, rather than exploration, method.

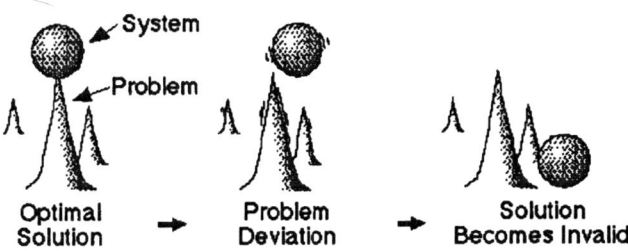

Figure 1.10. An optimal model (after Hale, 1996).

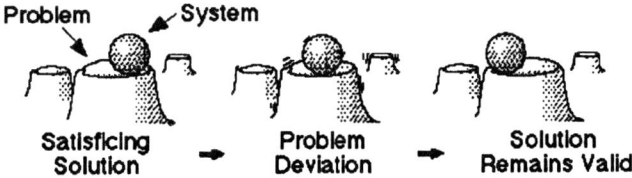

Figure 1.11. A satisficing model using solution mesas (after Hale, 1996).

Gero describes search as "routine design". By routine he means non-creative or non-innovative design. For example, the structural space of A-36 Schedule 40 steel pipe is well defined, as is the behavior space for compression loads on elements made with these sections. It can be considered "routine design" to search for a pipe size which results in a certain stress and stability behavior under a given condition. Exploration, on the other hand, might entail the expansion or contraction of the structure space to include other materials or other cross-sections. Gero describes exploration as "non-routine design" (Gero, 1994, p. 318). From this discussion it can be seen that search, and design methods that employ search techniques, are more applicable to later design phases where more decisions have been made, and more variables are set. Likewise, exploration is more appropriate to early design phases where the design state spaces are still under formation and likely to change. An IGDT is targeted for use in early design phases, and, therefore, has been developed to take advantage of exploration techniques.

1.2.3.5 Emergence in Design

Emergence is generally used to describe the order which a system or group of individuals can posses even though no single component or individual possesses this order. It has been referred to as order out of chaos (Holland, 1975). Commonly cited examples include ant colonies and the swarming of birds or insects (Johnson, 2004).

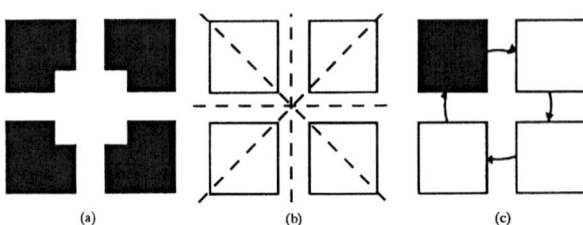

Figure 1.12. An example of emergence of form: (a) The four black 'L' shaped polygons are explicitly depicted. A white square is seen to emerge in the center of the figure. (b) Axes of symmetry can be seen as emergent - shown by dashed lines. (c) Rotational symmetry, shown by arrows, can also be seen as an emergent property. (from Gero & Jun, 1995)

Emergence, with respect to the relation of form to design, refers to the process of making implicit forms or shapes explicit. It is often seen as a way to expand the design space by suggesting new forms. In this sense it can be regarded as the opposite of fixation. Where fixation can result in a mental block, emergence can be a source of new conceptual directions. John S. Gero and his associates of the Key Centre of Design Computing at the University of Sydney have also contributed to the concept of emergent solutions found with the aid of GA's (Gero, 1997) (Gero & Ding, 1997) (Gero & Jun, 1995) (Jun & Gero, 1997).

The concept of emergence as applied to design of form plays an important roll in creativity. Designers often praise the value of sketching in developing new formal relationships. This is the traditional way in which forms can emerge in conceptual design.

Gordon (1961) describes several mechanisms for "making the familiar strange" as a way of stimulating creativity in problem solving. It is difficult for new ideas to emerge as long as one remains fixated on one solution.

There are two points concerning the emergence of ideas in design, and the visualization of those ideas which should be made.

- Old ideas are replaced only by new ideas.
- New, undeveloped ideas seldom replace older, well developed ideas.

The first point means simply that there is a natural reluctance to cast off a solution, even if it is not a very good solution, if there is no better possibility in the offing. It is so uncomfortable for a designer to be totally without a solution, that a poor solution will more likely be modified to work, rather than to be cast off leaving no replacement at hand. Kuhn makes this same observation with regards to science in general.

> Once it has achieved the status of paradigm, a scientific theory is declared invalid only if an alternative candidate is available to take its place. No process yet disclosed by the historical study of scientific development at all resembles the methodological stereotype of falsification by direct comparison with nature. That remark does not mean that scientists do not reject scientific theories, or that experience and experiment are not essential to the process in which they do so. But it does mean - what will ultimately be a central point - that the act of judgment that leads scientists to reject a previously accepted theory is always based upon more than a comparison of that theory with the

world. The decision to reject one paradigm is always simultaneously the decision to accept another, and the judgment leading to that decision involves the comparison of both paradigms with nature *and* with each other. (Kuhn, 1962 <1996 ed., p. 77>)

Kuhn's observations of the way scientists treat paradigms, or schools of thought, can be applied directly to the way in which designers will treat a design. One does not normally consider a design solution independent of other possible or previous solutions. This is an important concept applied in IGDT's, and lacking in typical design analysis programs. Because an IGDT offers several different solutions for the designer to compare and consider at any one time, it better allows for movement away from one solution toward another. On the other hand, analysis programs which may point out short comings in a particular solution or offer only one alternative, make it very difficult for the designer to choose another way or abandon the single solution offered by the program. As Kuhn stated, straight forward comparison with the facts of nature are not enough to allow change, an alternative has to be offered as well.

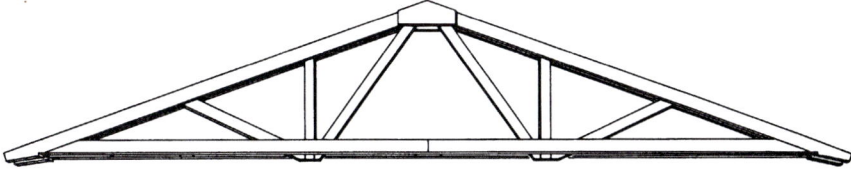

Figure 1.13. A Detail Instance of a Wood Truss (Schmitt, 1962).

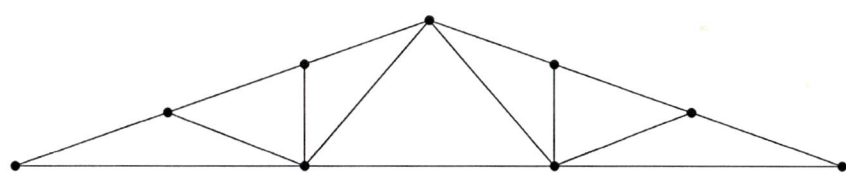

Figure 1.14. A More Ambiguous and Creatively Suggestive Line Drawing of a Truss with the Same Geometry as Figure 1.13.

The second point is, that a new solution is more likely to emerge from a less developed representation of a design. This is similar to the point made in Section 1.2.3.2., about too much knowledge too early in the design process which can stifle creativity, and limit the end solution. Although solutions are necessary to allow further speculation and new ideas, needlessly over detailing of the solutions (too much knowledge) detracts from the understanding of the basic idea and hampers the emergence of new solutions. For example, the detailed drawing in Figure 1.13 directs the interpretation to one specific instance, where as, the less distinct, more ambiguous line drawing in Figure 1.14 allows for several interpretations to emerge. The level of graphic detail in Figure 1.14 is at a more appropriate knowledge level for early conceptual design.

1.2.3.6 Lateral Thinking in Design

The term "lateral thinking" was coined by Edward de Bono in his book *Lateral Thinking for Management: a handbook for creativity* (de Bono, 1971). In this book de Bono explains the difference between lateral thinking and the more traditional vertical thinking, and builds an argument for the application of lateral thinking in conceptual design. De Bono defines vertical thinking as the traditional, logical, step-by-step approach to problem solving, where each step logically follows and builds on the preceding step, while expressing in and of itself a truth. Also known as the Socratic Method, vertical thinking was described by Plato, and has been the primary means of problem solving since that time.

De Bono does not argue the value or correctness of vertical thinking, but suggests that it is not always adequate, particularly when new or creative solutions are sought. De Bono is very pragmatic in his approach to problem solving. He is interested in arriving at a better solution. How this is achieved, through logic or illogical methods, is not important. If a solution is truly better, it will always be possible to show, after the fact, how one might have found the solution using logical, vertical thinking alone. But de Bono points out that first the new, creative solutions must be found, and that is where the application of lateral thinking can offer an advantage. The reason that lateral thinking is able to find creative solutions where vertical thinking fails, lies in the patterning nature of the human brain, and how it structures information. This patterning nature is trained by us to follow logical patterns, and will do so even if they do not lead to the best solutions. It can be very difficult to break out of a pattern of thinking and find new paths to a solution. This same phenomena is termed "fixation" in Gestalt theory, and is discussed in Section 1.2.3.3. Once started in a pattern, the mind will always favor continuing in the same pattern rather than changing over to a new pattern. De Bono proposes with lateral thinking several techniques which can aid the designer in finding and following new patterns.

> Lateral thinking is concerned with change - with the escape from old ideas and the generation of new ones. Lateral thinking involves two basic processes:
> 1. Escape
> 2. Provocation

(de Bono, 1971, p. 47-48).

An IGDT is particularly well suited to fulfill both of these criteria. By generating new solutions with minimal directional bias from the designer, it provides an *escape* from the designer's own preconceptions of possible solutions. Also by offering multiple solutions with a significant difference, the IGDT is more *provocative* than tools which offer a single best solution. These two points, defined so succinctly by de Bono, provide the key reason why an IGDT can function as a *design* aid and why traditional analysis tools cannot.

1.3 Tools for Design

This Section gives a broad overview of the types of design tools that have been developed and how they are applied. The scope has not been limited to either computational tools or computer based tools. Therefore, in this more inclusive dichotomy, design tools have been first divided into two categories:

- Non-computational tools
- Computational tools

In looking particularly at the non-computational tools, it is interesting to notice that the goal of the tool is not so much to provide the solution, as to provide the stimulus that leads the user to discover the solution. The IGDT is of course basically computational, but by including user selection, it is able to incorporate many of the aspects of the non-computational tools which strive to stimulate the creativity of the user.

1.3.1 Non-computational Design Tools

Non-computational design tools means those aids to design that are primarily not mathematically based. With current computational technology this might be understood as non-computer based, but this is not precisely what is meant. Computers can be used as a medium for displaying graphic images which offer no particular difference or advantage to the same graphic images published in a book. Although images may appear on a screen with the aid of sophisticated computational programming, they may otherwise remain basically non-computational, and be viewed and utilized in a basically non-computational way. The significance of this distinction in the context of an IGDT is that this same delineation of computational and non-computational seems to separate conceptual or early design tools from detailed or later design development tools. That is, traditionally most conceptual design tools are non-computational, where as most design development tools are computational. An IGDT is in this sense unique as a computational, conceptual design tool.

1.3.1.1 Text Based Tools

The most obvious text based tool is a book. Books are traditionally used to store knowledge specific to a field. The difficulty is having the right level of knowledge on the right topics available when needed. The level of knowledge recorded in books typically tends to be at too high a level, or too detailed, for conceptual design. This can lead to a knowledge overdose as described in Section 1.2.3.2.

There are several word games that can be employed as operational mechanisms to enhance creative thinking. Herbert Crovitz describes in *Galton's Walk* (1970), a relational-algorithm that can be used as a tool for exploring solutions to a variety of problems. Crovitz uses for his tool Ogden's Basic English, which is a subset of the essential 850 words needed to describe most common situations. With the premise that "action solves problems" (Crovitz, 1970, p. 96) Crovitz chooses the 42 words from Ogden's vocabulary which are able to make an elementary statement of action in the form "Take one thing ___ another thing". From Ogden's list the only words that can possibly fill in the blank are:

about	at	for	of	round	to
across	because	from	off	still	under
after	before	if	on	so	up
against	between	in	opposite	then	when
among	but	near	or	though	where
and	by	not	out	through	while
as	down	now	over	till	with

As examples of the technique, Crovitz uses the relational-algorithm to solve a series of problems proposed by Karl Duncker.

> Given a human being with an inoperable stomach tumor, and rays which destroy organic tissue at sufficient intensity, by what procedure can one free him of the tumor by these rays and at the same time avoid destroying the healthy tissue which surrounds it? (Duncker, 1945, p. 1)

Crovitz's relational-algorithm can only compare two things at a time, therefore, he regards first the possibilities of two rays and then the body and a ray.

> Take a ray about a ray.
> Take a ray across a ray.
> Take a ray after a ray.
> Take a ray against a ray.
> Take a ray among a ray.
> &c.

and for the case of the body and the ray:
> Take the body about a ray.
> Take the body across a ray.
> Take the body after a ray.
> Take the body against a ray.
> Take the body among a ray.
> &c.

Crovitz is able in this case to show solutions in both sets. "Take a ray across a ray" would represent a solution where individual rays were not of sufficient strength to destroy tissue, but when added by crossing would destroy the tissue at the point of intersection. In the second set, "Take the body round the ray" could be interpreted as rotating the body about the tumor so that the focus of the ray would be that center.

Crovitz's relational-algorithm can be an effective problem solving tool, but it is limited as a conceptual design tool to what can be expressed as action. Many design considerations do not fall into this set: aesthetics, meaning, expression, etc. Also, except for the possible imagery conveyed by words, it can suggest no form or space.

William J. J. Gordon, the creator of Synectics, also employed word based tools as "operational mechanisms" to "make the familiar strange" (Gordon, 1961). Gordon suggests that through the use of metaphor and punning, designers can expand their field of consideration by finding new combinations of words and phrases which, though they may initially seem irrelevant, can lend a new and profitable perspective to a problem.

> Neither logic as a system nor computer oriented "science" is capable of the reaches of metaphoric and analogic relevance which the creative imagination can develop in its search for forms. (Gordon, 1961, p. 130)

A. F. Osborn's Brainstorming techniques (Osborn, 1963) offer another well known example of the use of words to generate new ideas. In Brainstorming, words or short phrases are generated spontaneously, and with deferred judgment, by a group of individuals. The idea fragments, thus generated, are recorded for the group, and provide a sampling of solution spaces that can be used to suggest further consideration.

Although an IGDT is unable to draw associations between its autogenerated design forms and meaning, it can be seen to operate with a similar effect as Osborn's brainstorming, using images rather than words. The design forms generated by an IGDT do defer judgment until selection is performed on the population of solutions. In this way an IGDT, like brainstorming, fosters sampling of a wider portion of the solution space, and thus promotes exploration.

1.3.1.2 Graphic Based Tools

Many authors and designers have discussed the importance of sketching as a means of exploring potential concepts - (Broadbent, 1973), (Crowe & Laseau, 1984), (Antoniades, 1990). In her book, *Drawing on the Right Side of the Brain*, Edwards (1979) suggests that since drawing and creative thinking are both right brain hemisphere centered activities, they enhance one another. She relates her observations of how speaking (a left brain activity) is disruptive to students trying to draw, whereas music (a right brain activity) is not. Her premise is that since image processing and creative thinking are both taken to be right brain centered activities, they tend to be complimentary in a way that analytic analysis is not. The neurologist and researcher, James Austin, backs up much of Edwards observation about brain activity. He describes the respective activity of the two halves of the brain as follows:

> They do not behave as identical twins, nor is one the mirror image of the other. The inconstant wind of creativity fans different coals on the two sides of the brain. ... Our left cerebral hemisphere "thinks" in verbal, auditory terms, is good at translating symbols, including those of mathematics as well as language, and works best when analyzing a sequence of details. ... In contrast our right hemisphere "thinks" in visual, nonverbal terms, particularly in terms involving complex spatial relationships, and specializes in three dimensional depth perception. It also recognizes structural similarities, and works best in Gestalt: that is drawing conclusions based on a grasp of the total (visual) picture. It will instantly recognize a friend's face in a crowd, or perceive that a large skeleton key will fit a keyhole that has only a certain size and shape. Being adept at incidental memorization and the more musically gifted of the two, it may also prompt us to hum a long forgotten tune in an evocative surrounding. (Austin, 1978, p.138)

Certainly many examples can be cited in the writing and sketching of great designers, that would bear out the importance of graphic input to the design process: Le Corbusier (Corbusier, 1958), (Sekler & Curtis, 1978); Frank Lloyd Wright (Wright, 1943), (Storrer, 1993); Alvar Aalto (Schildt, 1989).

> Architectural designers often browse collections of images as they design, and the designing frequently involves drawing, copying, tracing, transforming and incorporating reference forms. Architects and design instructors encourage students to use visual references in developing design form. (Do & Gross, 1995)

Graphic based design tools are more common in practice than other forms of design tools. They cover a wide spectra, from the abstract generality of form types, to the specific case study. Graphic diagrams can be used to represent both structure and behavior, and as Alexander points out, truly constructive diagrams form a bridge between the two.

> The constructive diagram can describe the context, and it can describe the form. It offers us a way of probing the context, and a way of searching for form. (Alexander, 1967, p. 92)

In other words, they can represent a function which maps one to the other. In his method of decomposition, Alexander uses graphic diagrams to represent the subsets of the behavior space. Guided by linkages provided by decomposition he seeks an arrangement of the diagrams which indicates a structure (form).

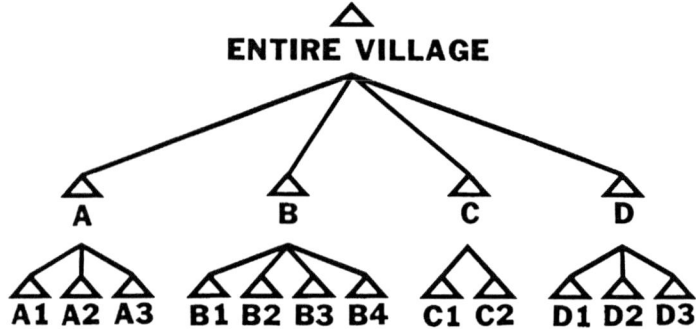

Figure 1.15. Alexander's tree diagram representing the hierarchy of subsets comprising the behavior space. (Alexander, 1967, p.151)

In his book, *Notes on the Synthesis of Form*, he develops an example of the method applied to the design of a village. Figure 1.15 shows Alexander's village represented with a tree diagram to represent the hierarchy of the subsets in the behavior space. Figure 1.16 shows the same village represented using graphic diagrams devised by Alexander which demonstrate a linkage between form and behavior.

Where as Alexander's use of graphics is fairly abstract, other graphic design aids can be more literal or form oriented. One aspect of the many Architectural magazines and journals which are published, is to stimulate the designer with images which communicate an established vocabulary of form.

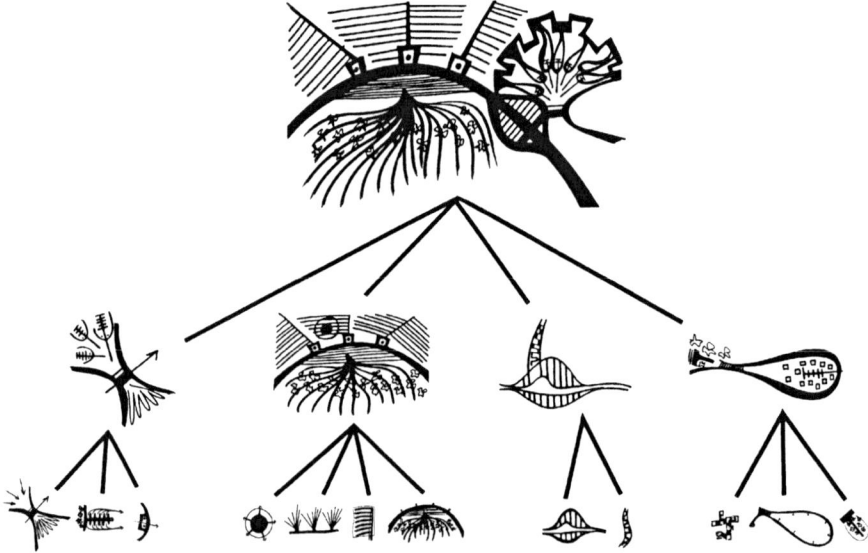

Figure 1.16. Alexander's graphic diagram representing simultaneously behavior and structure. (Alexander, 1967, p.153)

Associative connections are often generated by pictorial analogies, by a combination of images that belong to the observer's repertoire formed from personal experience. When stimulated by an actual or imaginary picture, they interact mutually and eventually lead to unexpected, but in a queer sense, related concepts. (Moro, 1996, p.24)

Illustrations showing the scope of forms which belong to a certain class of structures, aid the designer in visually understanding the attributes that belong to that class. Figure 1.17 shows an example of forms appropriate to the class of shell structures that can be constructed using pneumatic formwork. While no specific instance is intended, the designer can begin to recognize the range, as well as general class attributes, through comparison of the images.

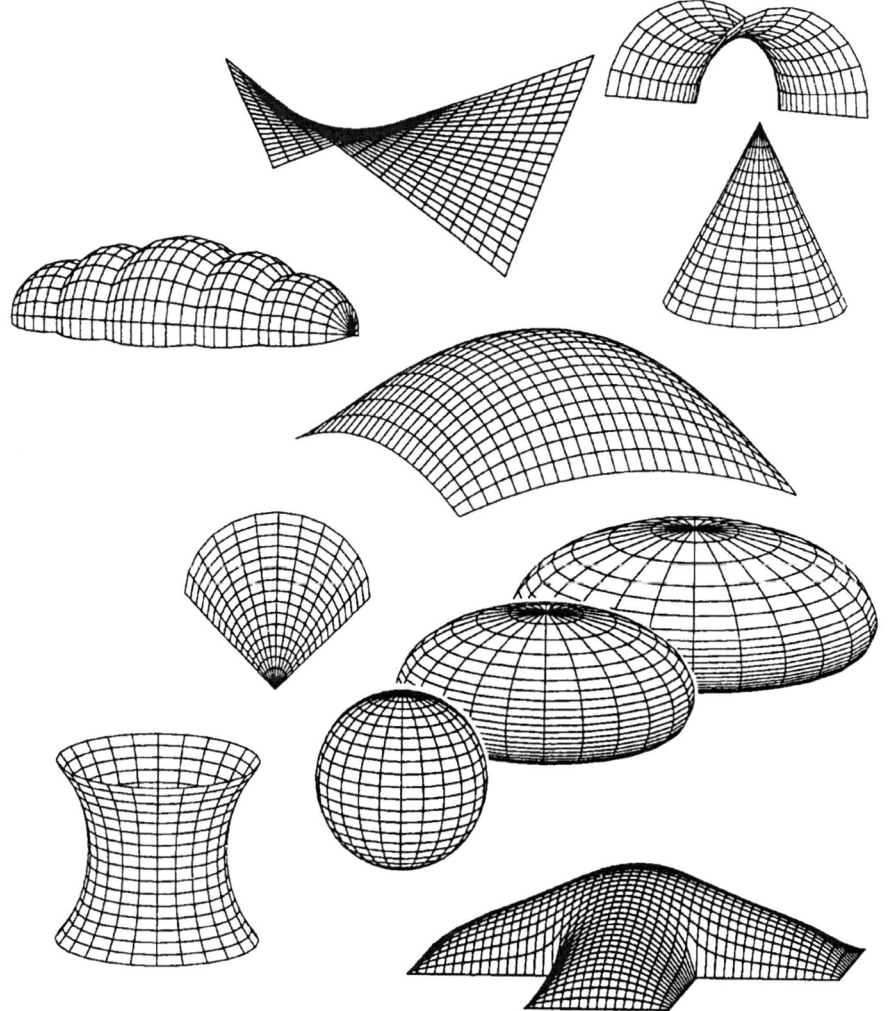

Figure 1.17. Examples of the Range of Structural Shells Which Can Be Pneumatically Formed. (Sobek, 1987, p. 37).

By simplifying the depiction of graphics, as in Figure 1.18, specific attributes can be made more apparent - in this case the stressed membrane form. To convey information about other characteristics, for instance behavior under light, more detailed graphics are required. While Figures 1.18 through 1.20 are all examples of the exact same structure, they indicate the range of information that can be conveyed with different graphic techniques.

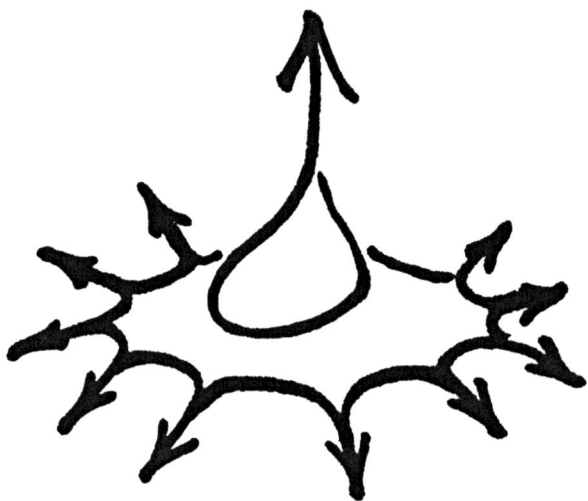

Figure 1.18.   Institute for Lightweight Structures: Concept Sketch.

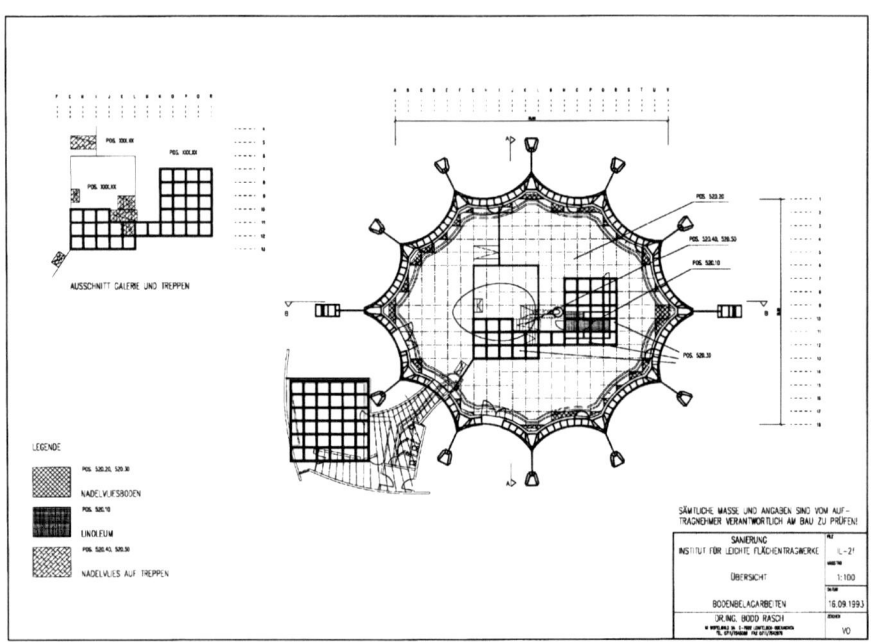

Figure 1.19.   Institute for Lightweight Structures: Architectural Plan Drawing (from 1993 restoration).

Figure 1.20. Institute for Lightweight Structures: Aerial Photograph (IL photo archive).

The limitation inherent with each of these graphic aids (with the possible exception of Alexander's which are intended to be generated by the designer as working tools) is that they tend to be static and non-interactive. A collection of pneumatic forms may help educate the designer to the vocabulary of the type, but as a static image it can not be used very effectively to explore new forms that belong to that same type. An IGDT combines the ability to display for the designer families of a structural type with the ability to explore with the designer new design spaces.

1.3.1.3 Model Based Tools

Physical models have long proven to be valuable design aids in exploring structural forms which meet certain behavior criteria. A well known example is Antoni Gaudí's use of funicular models to explore the structure of his masonry architecture. Figure 1.21 shows one of these models as reconstructed at the Institute for Lightweight Structures at Stuttgart University (Tomlow, 1989). With such models Gaudí was able to find a mapping for a specific behavior (compression stress under gravity load) to a specific structure in the form of the model.

Models, like those of Gaudí, that map behavior to structure are often called *form finding* models. Another example would be a three dimensional soap film model. Soap films can be used to find a minimum surface defined by a set of boundaries. This surface not only exhibits the behavior of economizing material, but also is characterized by a homogeneous, membrane stress state. Figure 1.23 shows a soap film model which approximates the structure described in Figures 1.18 though 1.20.

Figure 1.21. Reconstruction of Gaudí's funicular model of the church for Colonia Güell. (IL photo archive).

Another form of a soap film model has been applied to the solution of two dimensional Steiner problems, i.e., to find the shortest network of links between a set of points in a plane. *Minimal net* (or network) models can be used to explore routing or connection problems. In 1965, a device was constructed at the Institute for Lightweight Structures (IL) in Stuttgart based on work there of the preceding year, to generate and record minimal nets using soap films. Figure 1.22 shows an example of a minimal net generated with this device.

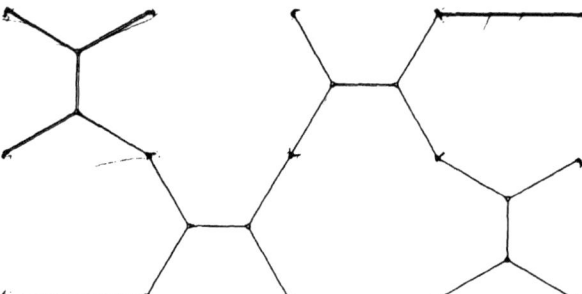

Figure 1.22. An example of a minimal net model based on a square grid. (Burkhardt, IL1, 1969)

Architectural models are often used to study a variety of physical parameters such as lighting, color and texture effects, climate control, etc. Dr. Alton J. DeLong, behavioral scientist at the University of Tennessee, has used large scale, transformable, architectural models to dynamically model space in response to human social behavior criteria (DeLong, 1981). De Long has shown that the dimension of time is governed by a scale factor in a similar way that other dimensional variables are scaled. This accounts for the faster assimilation of information that is experienced in a scaled space.

Physical models, because they invite participation on multiple sensory levels - visual, tactile, spatial, responsive, etc. - can be extremely good design exploration tools. Unfortunately, they tend to demand some skill in their creation and use, and a considerable investment of time to produce. The same physical nature that makes them so vivid as design aids, also limits their application by demanding that they be physically built. A major attraction to the virtual modeling of computers, is the ease and rapidity with which digital models can be altered. This rapidity allows computer design tools like an IGDT to explore a much wider design space than could be afforded timewise with physical models.

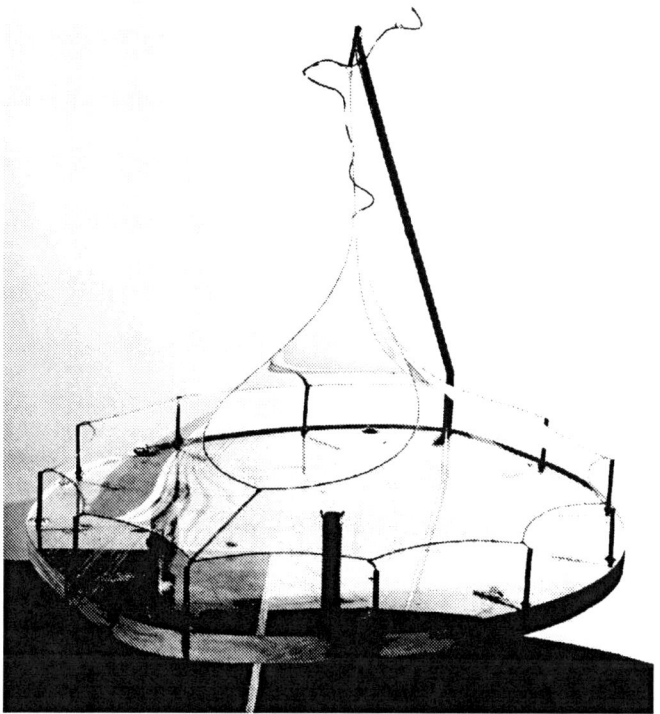

Figure 1.23. Soap film model of the Institute for Lightweight Structures. (IL photo archive)

1.3.2 Computational Analysis *versus* Design

In recent decades computer based tools have been developed for a variety of fields. Although originally applied to areas of computationally intensive *analysis*, with the ever increasing size and speed of processors, attempts have been made to develop *design* oriented applications as well. In the fields of architecture and engineering, the requirements of design tools are somewhat different from those of analysis tools. The analysis process can usually be executed in a regular, predetermined sequence of steps. The sequence may be iterative or have various paths based on the particulars of the problem, but there remains one "correct" path, using a chosen method to solve a stated problem. For example, for a given structural member, with a given load, a particular analysis will yield one solution for the stresses at some point. The analysis is composed of a sequence of prescribed steps which lead to the solution. Each time the analysis procedure is executed it leads to the same result. To the degree that different individuals or different machines show a discrepancy in their results, it is assumed that some error is present. In fact, an analysis is usually verified by showing consistency with the results obtained in another way. This, of course, is not to say that various approximate methods will not yield solutions with differing degrees of accuracy. But the fundamental premise remains, that for a specific load state, there will exist one specific stress state.

In contrast, the design process is not expected to consistently yield the same result. Although a designer may follow a sequence of steps, the steps are not self contained, but influenced by factors outside the process itself (the unique background of the designer, stimuli of the environment, *Zeitgeist*, etc.). For a given design problem with a given set of parameters, it is certainly not expected that any two designers will come to the same solution. One need only look at the results of any design competition to see the variety of solutions that can be proposed. If a competition for a bridge or building resulted in every entry being identical, the competition would be considered a failure. Design implies creative thinking, and creative thinking does not fit a predetermined set of serial steps.

Another way to see this is with reference to the distinction between search and exploration made earlier in Section 1.2.3.4. In this sense, analysis is a search process, while design is exploration. Traditional analysis tools can only search one behavior space for a set of variables that will result in a desired structure. But choosing or defining the behavior space is part of design, and requires special attention in the development of design tools.

With the advent of parallel processing it was supposed by some, that computers would be more capable in the area of design as compared with the earlier serial processing machines. After all, being able to hold two ideas simultaneously is a prerequisite to using metaphors, and metaphors are closely linked to creative thinking (Gordon, 1961; Prince, 1970). But, although parallel processing can shorten computing time, it has not had much impact on enhancing the design capabilities of computer tools. Parallel processing alone does not automatically ensure a shift from search techniques to exploration techniques. Or as de Bono might term it, design involves not so much parallel, vertical thinking as "lateral thinking"

> Vertical thinking uses information for its meaning. Lateral thinking uses information for its effect in setting off new ideas. Vertical thinking is analytical. Lateral thinking is provocative. Vertical thinking is interested in where an idea comes from: this is the backward use of information. Lateral thinking is interested in where an idea leads to: this is the forward use of information. (de Bono, 1971, p.9)

Even work in the area of artificial intelligence has difficulty in showing capabilities of lateral thinking. For this reason computerized tools have found more success in the area of analysis, which is more easily described in a vertical procedure, than they have in design.

1.3.2.1 Computational Analysis Tools Used for Design

More powerful optimization programs might be capable of searching optimal geometries or even optimal topologies and optimal geometries. But the focus of such tools is still on the analysis phase of design. These tools analyze a given set of criteria to find one, single solution. As tools, they obscure to the user alternate directions, i.e., alternate behavior state spaces. Such tools tend to imply, with a final solution, an end to the search process. Therefore, they cannot function effectively as idea generators, or be used to aid in the formulation of the concepts that direct their own optimization. In fact, by offering the user a single solution, such tools tend to limit creative thinking on the part of the designer.

In terms of conventional mathematical methods (e.g., Linear Programming), a solution to a problem is described in terms of one or more *variables* which are limited by *constraints*. The solution is optimized with respect to an *objective*. For example, if the problem is to optimize the profile of a simple beam, the objective might be to minimize weight. Variables could be the sectional geometry – perhaps width and depth. The constraints might be stress levels in the material or deflections.

If there is only one variable in the problem, simple calculus methods can be used to find maxima and minima bounded by the constraints. If more variables are needed to describe the solution, methods such as Linear Programming would be needed. Or if there are several objectives (one might want to minimize weight and maximize modal frequency) multi-objective or multi-criterion methods are needed. In the latter case, a set of solutions can be found (a Pareto set), but usually the goal is still to find a single best.

Another mathematical approach used in the optimization of continuum structures is the homogeneous method. In this approach the design field is subdivided into a mesh of cells which can be varied in density depending on relative stress levels caused by loadings. In different variations this is called Topology or Shape Optimization (Bendsøe & Kikuchi, 1993).

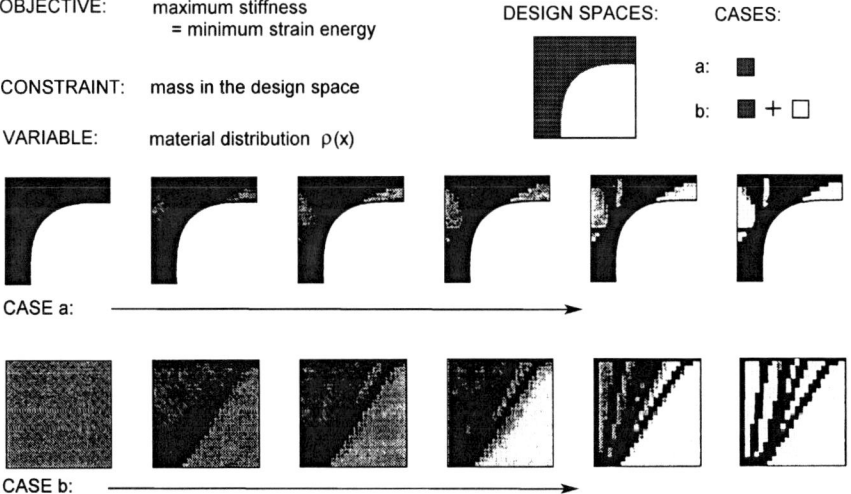

Figure 1.25. Two examples of topology optimization using CARAT. Both examples depict a bridge pier with the same material and loading conditions. In case 'a' the design space is restricted to the shaded area, while case 'b' is allowed to utilize the full rectangular envelope. (Ramm et al., 1997, p.211)

An example of a program which takes advantage of this type of shape optimization is CARAT, developed at Stuttgart University, Institute for Structural Mechanics (Ramm et al., 1997). Figure 1.25 shows two examples of topology optimization using CARAT.

Any of the above optimization methods can be very effective as an analysis tool, either for skeletal or continuum structures. But for use in design exploration, they would generally require intervention on the part of the user (e.g., repeatedly altering some parameters – constraints or variables) in order to generate different solutions.

In concept this is very different from the design orientation of the IGDT. In focusing on one single solution, analysis tools only made visible a minimum of the solution space to the designer. In offering the designer only one solution (or even many solutions, one at a time) the tendency is to limit consideration of alternate paths. Rather than being aided, creativity tends to be stifled.

Also, traditional methods are limited in the types of objective functions they can work with. Linear Programming tends to imply linear objective functions, and functions are more easily differentiable if they are continuous. Although there are programming methods which can deal with non-linear functions, and make discrete functions more workable, many of the designer's criteria do not lend themselves to numeric functions well at all. This is particularly true in cases where seeing possible solutions helps to formulate the design parameters. In early exploration phases of design, this is often the case.

Evolutionary Algorithms are of course also mathematical methods. Although the methods using Genetic Algorithms are well suited for user interaction, most implementations to date have imitated traditional analysis programs in their execution. With no designer interaction and a goal of one 'best' solution, the same limitations ultimately restrict the effectiveness of genetic methods as other numerical methods. Several authors have in used Genetic Algorithms to optimize structural systems (Adeli, 1993; Bouzy, 1995; Cai, 1995; Cheng, 1992; Coello Coello, 1994; Galante, 1996; Grierson, 1993; Hajela, 1995; Höfler, 1976; Koumousis, 1994; Leite, 1995; Louis, 1995; Powell, 1993; Rajan, 1995; Rajeev, 1992; Ramasamy, 1996; Sugimoto, 1992; Wu, 1995; Xu, 1994; Yang, 1995; Zarubin, 1995). In most cases they duplicate efforts already achieved by other mathematical methods. In some cases the authors offer one advantage or the other. For example, GA's may be more suitable for optimizing non-linear functions (Powell, 1993), or optimization using discrete sizes of members (Galante, 1996). But in most cases, GA's are still being used in the same way that other mathematical methods are used. They are being used as analysis tools rather than design tools. A notable exception is provided by Emanuel Slaby (2003).

Another approach to consider which uses stochastic computation is the Multi-Objective Evolutionary Algorithms (MOEA) (Coello Coello, 2004). These belong to the field of Evolutionary Multi-Objective Optimization (EMOO). In concept a MOEA is fairly similar to the IGDT in that it seeks to find a 'good' solution to a complex problem. Also, the Pareto set found using a MOEA is similar to the multiple 'good' solutions found by the IGDT. But a MOEA requires explicit definitions of each of the objectives. The IGDT on the other hand, has the flexibility to use either a single objective or either explicit or implicit multi-objectives. The number of 'good' solutions returned by the IGDT is not limited by the number of objectives.

A Vector Evaluated Genetic Algorithm (VEGA) is another form of EMOO. First published by J. David Schaffer in 1984 (Schaffer, 1985) VEGA uses different sets of populations to find 'good' solutions to the different objectives, and then combines these solutions into one overall population that is acted on by the GA operators of mutation and crossover. This differs from the MOEA in that it is not finding Pareto solutions (good solutions to combinations of objectives) but seeks a good solution from a set of good performers to the individual objectives. Since the IGDT works on all objectives at once, it is more likely to find the good Pareto solutions.

1.3.2.2 Computer Aided Design (CAD)

Computer Aided Design (CAD) programs have evolved a great deal from early days of computer assisted drafting. In the current use of the tool, a CAD application is used to develop a virtual model of the designed structure. Once an digital model of the structure is in place, it can be subjected to various pertinent analyses, commonly including some form of finite element structural behavior analysis. Fully developed CAD packages generally include rendered visualization routines and software that can depict light and shadow effects. Modeling, analysis and visualization software combine to give an effective and efficient basis for design. John Abel, Professor of Structural Engineering at Cornell University, describes CAD as providing the designer with control over elements of model creation, analysis and design within a computer environment (Abel, 1997). CAD can then be used as an iterative design cycle in which the designer develops the virtual model of the structure through successive trials and modifications. Figure 1.24 shows the schematic implementation of this concept.

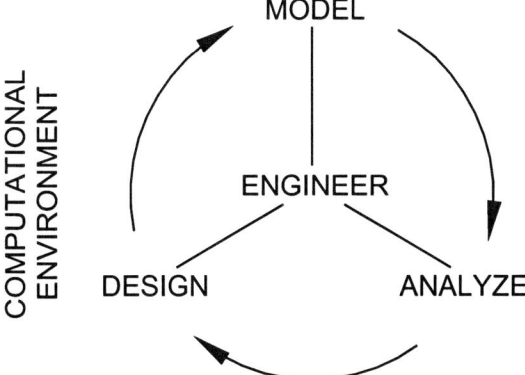

Figure 1.24. Schematic of the structural CAD process and its components. (after Abel, 1997, p. 190)

The Model, in Abel's CAD implementation, is an digital data base which includes a description of geometry, topology, material, properties, geometric properties, loads and support conditions, as well as other special quantifiable design attributes. This is basically implemented currently in several commercial packages (to one degree or another) in the form of Building Information Models or BIM's. These elements usually come in the form of parametric solids. Although differences do exist between architectural models and structural analysis models (Johnson, 2004), attempts are being made to include structural information for purposes of analysis in BIMs.

The Analysis component of a CAD package models the design as a meshed solid or other discrete pieces, which can be analyzed by finite element methods. The type of analysis can actually vary depending on the level of design detail, e.g., static, linear analysis may be adequate in initial design phases that may later require dynamic or non-linear final analysis. The necessary material and geometric properties can then be supplied by the BIM elements described above. Depending on the level of sophistication, the analysis phase may simply deliver structural behavior of the elements, or may include a form optimization.

The Design component of the cycle suggests changes to the CAD data base. This may be accomplished by a rule based system such as an expert system, or changes resulting from optimization methods applied at the level of material, member shape, geometry or topology. The information is made available to the design engineer in a way that supports his understanding of the choice options and consequences in a clear and timely fashion. The computing environment used to accomplish this interaction is some form of graphic user interface (GUI).

Führer's development of ExTraCAD (Führer, 1991) can be cited as an early example of this class of analysis aid. ExTraCAD provides a comfortably graphic shell for a more traditional finite element analysis (FEA) program. The CAD shell makes the logistics of using the FEA program easier and the output more readily understandable. The original CAD model can be readily modified based on the FEA results. But, conceptually the program remains analysis oriented, and as such tends to lead the designer strongly in one direction.

Although CAD has been a boon with regards to design development, the exploratory aspect of conceptual design can actually suffer in a CAD environment if not carefully considered. In a study made by Goel comparing the behavior of designers using MacDraw (a simple CAD program) with designers using paper and pencils, those working on machines tended more to take an initial idea and develop it throughout a session, whereas students working on paper were more likely to explore a series of different directions (Goel, 1992). This implies that using a CAD program for conceptual design can actually have a restrictive effect on a designer's creativity.

1.3.3 Computational Design Tools

In this book, the term "intelligent design tools" refers to design aids which are able to anticipate the designer's own preferences in the exploration of the design space. Mihai Nadin refers to this as "Computational Design".

> The strength of the human being, as a creative entity, is in anticipating, not in reacting to the outside world and its natural changes.
> Computational design is by its nature anticipatory, proactive. (Nadin, 1997, p. 53)

The ability "to anticipate" belongs to design in the way that "to react" belongs to analysis. In the development of design tools, this anticipatory nature is key. Traditional design tools anticipate some pre-programmed definition of good design, whereas intelligent design tools anticipate the designer's own concept of a specific design instance.

What qualifies a design tool as intelligent is the ability to respond to implicit directives of the user. That is, without explicit programming, the tool "understands" what is desired by the user, and is able to respond. This is accomplished in an IGDT by the use of Genetic Algorithms coupled with user selection as described in Section 1.3.4.

1.3.3.1 Development of Artificial Intelligence

The ability of machines to interact intelligently with humans and creatively solve problems has been dreamed of and debated since before the advent of the modern computer. Alan Turing, outlined arguments in support of and against thinking machines in his essay published in 1950 in *Mind*, "Computing Machinery and Intelligence". In the article he describes the Turing Test for intelligent behavior in the form of an "imitation game" played between a human interrogator, a human contender and a machine contender. In short, the interrogator has the task to determine which of the two contenders is the human, and which is the machine, by asking a series of questions. Turing's contention was, that if the interrogator is unable to distinguish accurately between the behavior of the machine and the behavior of the human, then the machine must be behaving like the human, viz. intelligently.

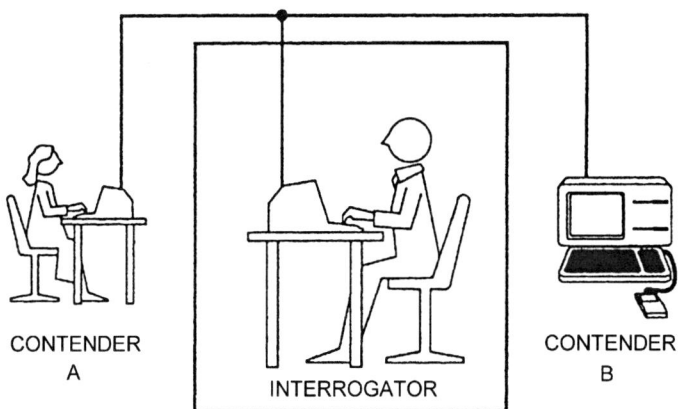

Figure 1.26. The Turing Test: the interrogator is given the task of determining which contender is machine and which is human. (after Luger & Stubblefield, 1989)

This simple test has several advantages that has made it very attractive as a means of assessing the success of an AI program.

- It gives a pragmatic and testable definition to intelligence, without entering a debate of what comprises intelligence.
- It avoids any reference to the inner processes used by either human or machine contender.
- It avoids any bias of physical attributes such as appearance, speech or actions and focuses on content of answers.

Turing believed that, "the problem is mainly one of programming" (Turing, 1950), and even estimated the amount of knowledge needed, and the amount time it would required to code that knowledge: "...about sixty workers, working steadily through the fifty years might accomplish the job" (Turing, 1950). Interestingly enough, those fifty years have now past, with a good deal more than sixty people engaged in AI programming, yet, except perhaps in a very limited sense, no machine has passed Turing's test.

Turing himself pointed to an alternative to direct programming that has also attracted much research effort over the decades, but has likewise failed to produce a winning candidate for the Turing Test.

> Instead of trying to produce a program to simulate the adult mind, why not rather try to produce one which simulates the child's? If this were then subjected to an appropriate course of education one would obtain the adult brain. Presumably the child brain is something like a notebook as one buys it from the stationer's. Rather little mechanism, and lots of blank sheets. (Mechanism and writing are from our point of view almost synonymous.) Our hope is that there is so little mechanism in the child brain that something like it can be easily programmed. The amount of work in the education we can assume, as a first approximation, to be much the same as for the human child. (Turing, 1950)

What one can conclude from the decades of work that have gone into Turing's second suggestion, is that the mechanisms of the organic brain - even a child's brain - are *deceptively* simple, and that far from being analogous to a blank notebook, the brain seems to come preprogrammed with a surprising amount of information and rules that have yet to be understood by the adult mind. (Pinker, 1995)

For several decades AI researchers followed Turing's direction, and tried to beat the "imitation game". Rather than to define intelligence in other ways, the attempt was made to simulate intelligence by mimicking human behavior. Two areas of human behavior that seemed most to exemplify intelligence were problem solving and related strategies in game playing. To that end, many programs were written that could play some strategy game or solve some type of problem as well as a human. Early success of such imitation programs lead many to believe that further degrees of intelligence could be accomplished by simply more of the same, viz. larger programs that defined more rules and supplied increasing amounts of knowledge that applied to those rules. This strategy is perhaps best expressed in the quest through the years to develop a machine that could better any human at the quintessential game of strategy, chess. In the 1968 screenplay *2001: A Space Odyssey* by Stanley Kubrick and Arthur C. Clarke, the superior intelligence of HAL, the ships computer is indicated by a scene in which it easily beats one of the ships officers in a game of chess. Roger Schank, AI researcher and expert in the field of natural language understanding, comments on the scene:

> The writers of 2001 made the same mistake that artificial intelligence (AI) researchers made about intelligent machines - a mistake that dates from the very beginning of AI research. They assumed that an entity that engages in intelligent actions is, therefore, intelligent. ... Early AI work relied on chess-playing programs as a kind of "quick hit." Success was relatively easy, and all of a sudden computers seemed pretty smart. The problem is that what early AI researchers took as evidence of being smart was illusory.
> (Schank, 1997, p. 174-175)

With the successful out matching of world champion chess player Garri Kasparow by IBM's Deep Blue in 1997, the AI objective of a machine chess player better than any human chess player was realized. Nonetheless, the AI goal of a machine capable of matching human intelligence seems today less plausible than it must have seemed to Turing's listeners in the 1950's.

Dr. John R. Searle, Berkeley professor and specialist in the philosophy of the mind and language, argued in his 1980 paper "Minds, Brains and Programs" against Turing's notion of "strong AI" (Searle, 1980). Searle develops an interesting analogy to AI programs in which a man, isolated in a room, follows a set of instructions which make it possible for him to answer questions in Chinese about a story, also in Chinese, by treating the Chinese glyphs strictly as symbols, that is, without any understanding of Chinese. This, Searle states, is analogous to the AI computer program which is all "syntax but no semantics" (Searle, 1980). No matter how much syntactical knowledge may be coded into a program, the program itself remains like the elaborate instructions given to the man in the isolation room. Because no *meaning* is attached to the instructions, even if the man in the room were to memorize the instructions, and even if the instructions were so well conceived that the man's answers in Chinese, to questions in Chinese, about the story in Chinese, were so flawless that any observer outside of the isolation room would have to assume that the man in the room must understand Chinese, despite all appearances, the man in the room would, indeed, have no understanding of Chinese at all. And so, any expectations that the man in the room would ever be able to go beyond the programmed instructions to, for example, start conversing freely in Chinese on topics other than those covered by the programmed instructions, would be totally impossible since the man has no understanding of Chinese. Therefore, by Searle's analogy, one can not expect any AI program to ever attain some threshold level of knowledge, which would enable the system to continue to learn and function with meaningful intelligence. Searle makes the following observation based on his analogy.

> ...could something think, understand, and so on *solely* in virtue of being a computer with the right sort of program? Could instantiating a program, the right program of course, by itself be sufficient condition of understanding? ... the answer ... is no. ... Because the formal symbol-manipulations by themselves don't have any intentionality; they are quite meaningless; they aren't even *symbol* manipulations, since the symbols don't symbolize anything. In the linguistic jargon, they have only a syntax but no semantics. Such intentionality as computers appear to have is solely in the minds of those who program them and those who use them, those who send in the input and those who interpret the output. (Searle, 1980, p.83)

Many AI researchers today no longer hold the expectation of Turing and other early AI researchers, that a machine could be developed with intelligence indistinguishable from human intelligence. Schank expresses the more recent view of many AI researchers as follows:

> Such machines will be local experts; that is, they will know a great deal about what they are supposed to know about and miserably little about anything else. They might, for example, know how to teach a given skill, but they will not be able to create a poem or play chess. They might be able to converse about the day's news to keep a user informed, but they won't know how to fly a rocketship. Or, they might be able to fly a rocket ship but not be able to identify George Washington. ... Current efforts in AI are focused on producing just such devices. It is unlikely that we will ever see a HAL. Although this realization may be evidence of a dream abandoned, it may foreshadow the development of "real" artificial intelligence.
> (Schank, 1997, p. 189)

The IGDT can be seen as belonging to Schank's description of "'real' artificial intelligence". It is conceived of as a tool to be an aid to and used by the designer. But in is not seen as a replacement for the designer. In the sections which follow some of the current AI techniques which have been applied specifically to the area of architectural engineering design are described.

1.3.3.2 Expert Systems

Expert systems have had a long development in the area of AI. During that development some of the initial expectations have changed over the years. Basically, expert systems all involve a quantity of knowledge in a specific domain coupled with some heuristic methods of accessing or applying that knowledge. In summary, the development of expert systems over the years has been concentrated in two general approaches (Stipp, 1995), (Brooks, 1989):

- Knowledge-Based Systems   (Top-down)
- Learning-Based Systems   (Bottom-up)

In the Knowledge-Based Systems approach, the *knowledge* is based in a series of rules which constitute a centralized resource for all decision making. Any decision made comes from this centralized *top* down. By contrast, in a Learning-Based System there are hierarchical layers of complexity which build from the *bottom* upward toward more complex systems. Each layer adds additional meaning or complexity to the more basic layer below it. The bottom then provides the most fundamental, functional intelligence for the system. Researchers in both groups postulate (at least originally) that there is some threshold of programmed information that will allow a system to learn further on its own, thus enabling it to bootstrap its way to higher levels of creative intelligence. As yet no one has achieved this threshold level of intelligence in either group.

A major proponent of the knowledge-based approach is Douglas Lenat. Since 1985 he has been attempting to develop a knowledge base of all essential human knowledge. Lenat hoped that his program, Cyc, would be able to attain a basic level of intelligence that would allow it to teach itself more complex concepts, and eventually be capable of spontaneously developing new theories or solutions which could span different areas of knowledge. But despite some $25 million in funding from the US Nation Security Agency, this basic level of intelligence was not achieved. Although Cyc can be used to solve specific problems, Lenat has not been able to get it to a level that might be considered creatively intelligent. Lenat's experience has been shared by many who have worked for years programming knowledge into expert systems. The initial promise of expert systems was, that by gaining enough basic knowledge, the system would eventually be able to creatively solve problems internally. This hope has not been fulfilled as expert and knowledge based systems remain limited by the specific knowledge and logic patterns for which they have been trained. The initially irrelevant or illogical information which forms a bridge to creative solutions, stands as a wall which prevents the system from making the creative leap.

In the learning-based approach to AI, the emphasis is on developing a system which can learn through sensory detection. The AI Lab at MIT under the direction of Rodney Brooks is a leader in this area. In this approach the emphasis is placed on defining knowledge collection mechanisms rather than specific instances of knowledge. Rather than attempting to encode the intelligence of an adult with the goal to learn, this group attempts to encode the sensory information gathering ability of a baby which teaches itself by interacting with an environment. The group at MIT has demonstrated some success with this approach in the form of robots that can move about and perform simple tasks like collecting coke cans (Kelly, 1995) or blow up land mines (Stipp, 1995). Brooks tries to build in intelligence from the bottom-up. By close coupling of sensors and actuators, the attempt is to replicate the brainless nervous system responses found in lower life forms. When a sufficient number of low level responses are established, it is postulated that an emergent intelligence will appear.

Simon likens this to the activity of an ant (Simon, 1969, <1996 ed., p. 52>). Although, for example, the path it takes in foraging for food may appear to be rather complex, on closer analysis, it can be seen as a series of simple responses to conditions it encounters.

The apparent complexity arises due to the environment, not the "program" which steers the ant.

1.3.3.3 Shape Grammars

Shape grammars are sets of rules which govern the arranging of shapes in space. Shapes can be either two dimensional Euclidean type geometric figures or more complex three dimensional solids. In addition shape labels can define non-geometric attributes, such as color or line weight, which are used to impart further meaning for the shape, e.g., usage, adjacency requirements, value, etc. Shapes can also be defined using parametric variables. The rules which form the grammar, are based on Boolean operations of union, intersection and difference coupled with logical statements of the form $\alpha \rightarrow \beta$, which implies if the shape α is found then replace it with the shape β. Rules can also include transformation operations of translation, rotation, reflection, scale or combinations there of. Transformations can be applied to a shape, denoted by $\tau(s)$ or applied to a shape label, denoted by $\tau(p)$.

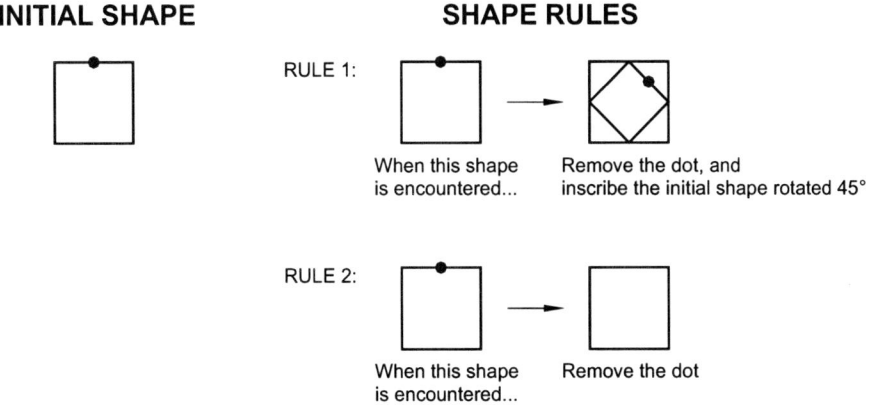

Figure 1.27. An example of a simple shape grammar derived from two rules. The • indicates a shape label. (Stiny, 1980, p. 348)

Figure 1.28. Generation of a shape using the shape grammar shown in Figure 1.27. (Stiny, 1980, p. 348)

A detailed description of the mechanics of shape grammars can be found in a series of articles by George Stiny, a major pioneer of the field of shape grammars and currently a Professor of Architecture at MIT. Stiny's article in *Environment and Planning B*,

"Introduction to Shape and Shape Grammars", defines much of the ground work for basic shape grammars (Stiny, 1980). Figures 1.27 and 1.28 show an example of the definition and application of a shape grammar by Stiny.

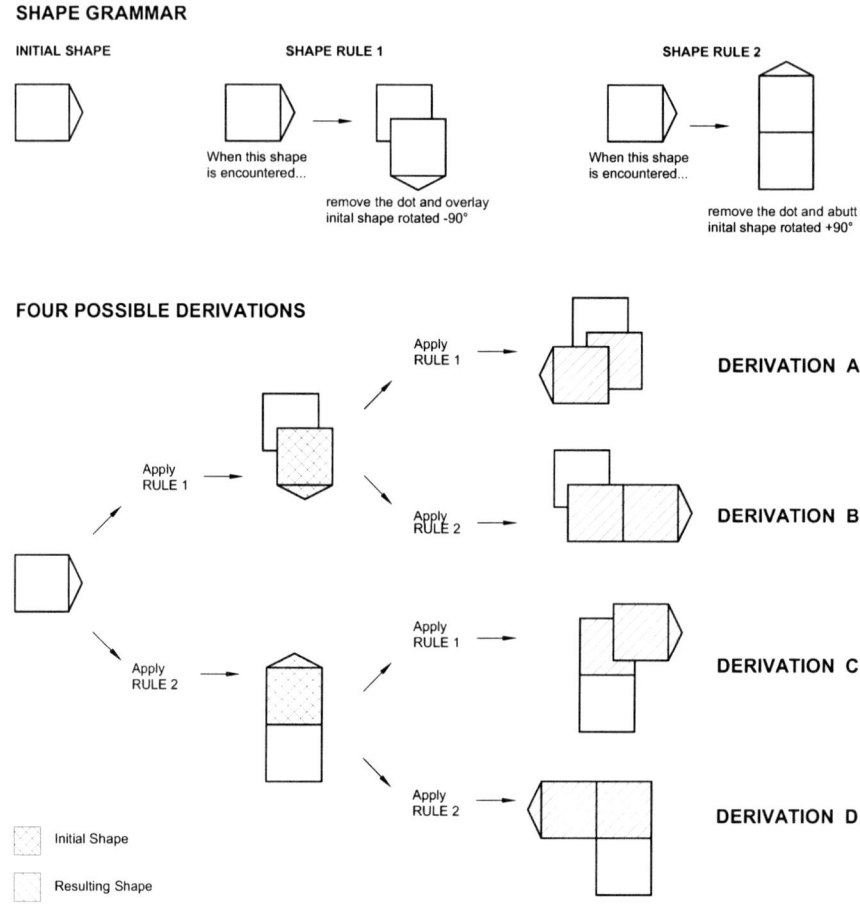

Figure 1.29. A nondeterministic shape grammar that allows a selection between two different rules to apply to a particular part of a design. (after Knight, 1999a)

Terry W. Knight, also in the Department of Architecture at MIT, has more recently added further definition to Stiny's basic shape grammar (Knight, 1999a; 1999b). Knight delineates six categories of shape grammars, based on restrictions applied to the rule orderings.
- basic grammar
- nondeterministic (ND) basic grammar
- sequential grammar
- deterministic grammar
- unrestricted grammar

Basic grammars, like Stiny's version discussed above, are linearly ordered set of addition type rules: $r_1, r_2, \ldots r_n$. Basic grammars are deterministic, meaning that there is only one possible outcome to the application of the grammar from a given start configuration. In the nondeterministic form, the set of rules is only partially ordered. This means that in some circumstances a choice between rules or their application can exist. This choice makes possible various outcomes. Figure 1.29 through Figure 1.31 show three examples of possible nondeterministic shape grammars. In each case different end results, or derivations, can be reached depending on choices made at each juncture.

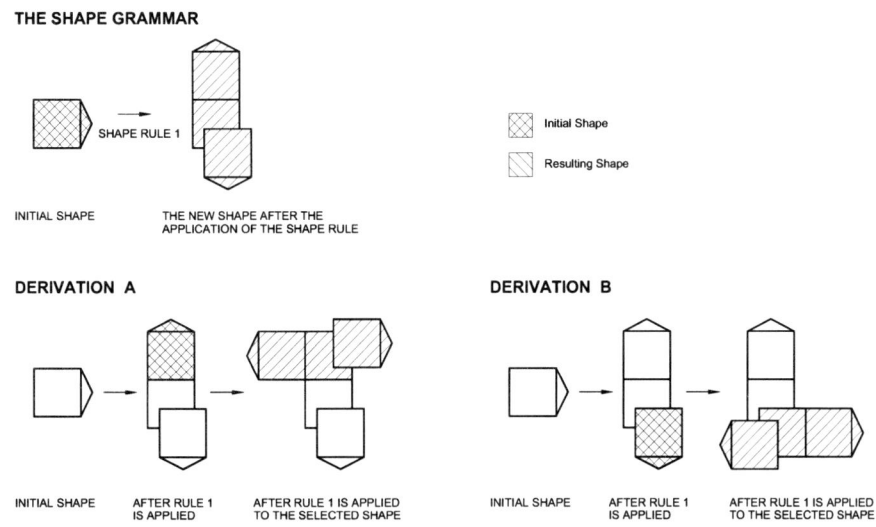

Figure 1.30. A nondeterministic shape grammar that allows a selection between different parts of a design to apply one rule. (after Knight, 1999a)

Several researchers have successfully applied shape grammars in the exploration of variety of pattern based design problems. Knight's work includes decorative furniture patterns (Knight, 1980), De Stijl paintings (Knight, 1989), and Greek pottery motifs (Knight 1986). Stiny has produced grammars for Chinese lattice designs (Stiny, 1977), and with Mitchell, Mughul gardens (Stiny & Mitchell, 1980) and grammars that reproduce Palladian villas (Stiny & Mitchell, 1978). Downing and Flemming have coded grammars which reproduce Bungalow style houses, while Koning and Eizenberg have successfully generated an array of Frank Lloyd Wright, prairie style houses (Koning & Eizenberg, 1981). Figure 1.32 shows examples of some of these applications. Figures 1.33 and 1.34 show in more detail the results of the Frank Lloyd Wright grammar from Koning and Eizenberg.

In observing the examples of shape grammar applications it is apparent the class of problems that is best suited for this method. Although potentially any conceivable form can be seen as being built up from constituent shapes, in practicality, shape grammars can only be written for patterns that are recognizable and can be defined by rules.

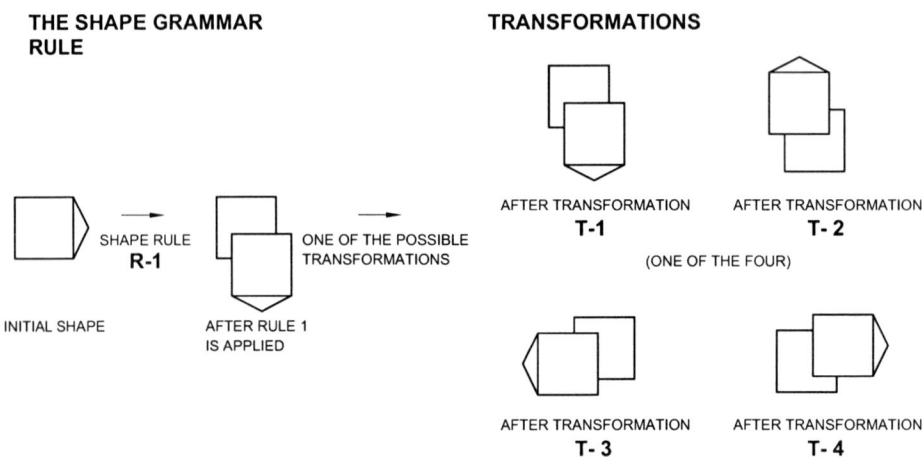

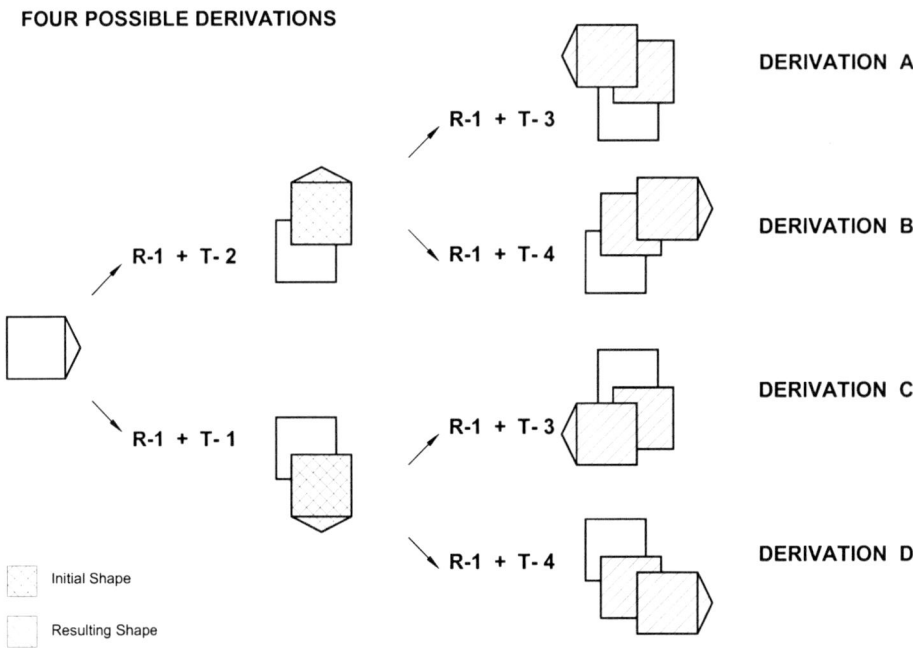

Figure 1.31. A nondeterministic shape grammar that allows a selection between different transformations under which one rule may apply to a particular part of a design. (after Knight, 1999a)

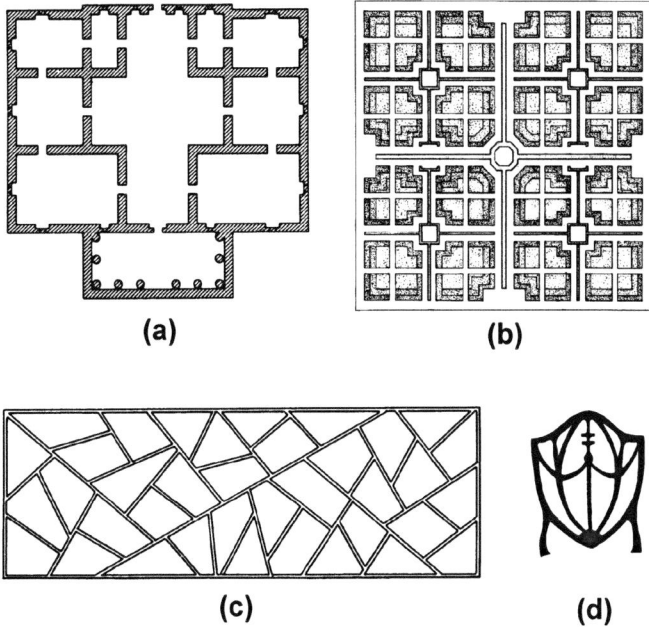

Figure 1.32. Examples or solutions derived by shape grammars. (a) Palladian villa plan by Stiny & Mitchell; (b) Mughul garden plan by Stiny & Mitchell; (c) Chinese lattice design by Stiny; (d) Hepplewhite-style chair by Knight

Figure 1.34. Perspective rendering of the "Stiny House" from Figure 1.33.

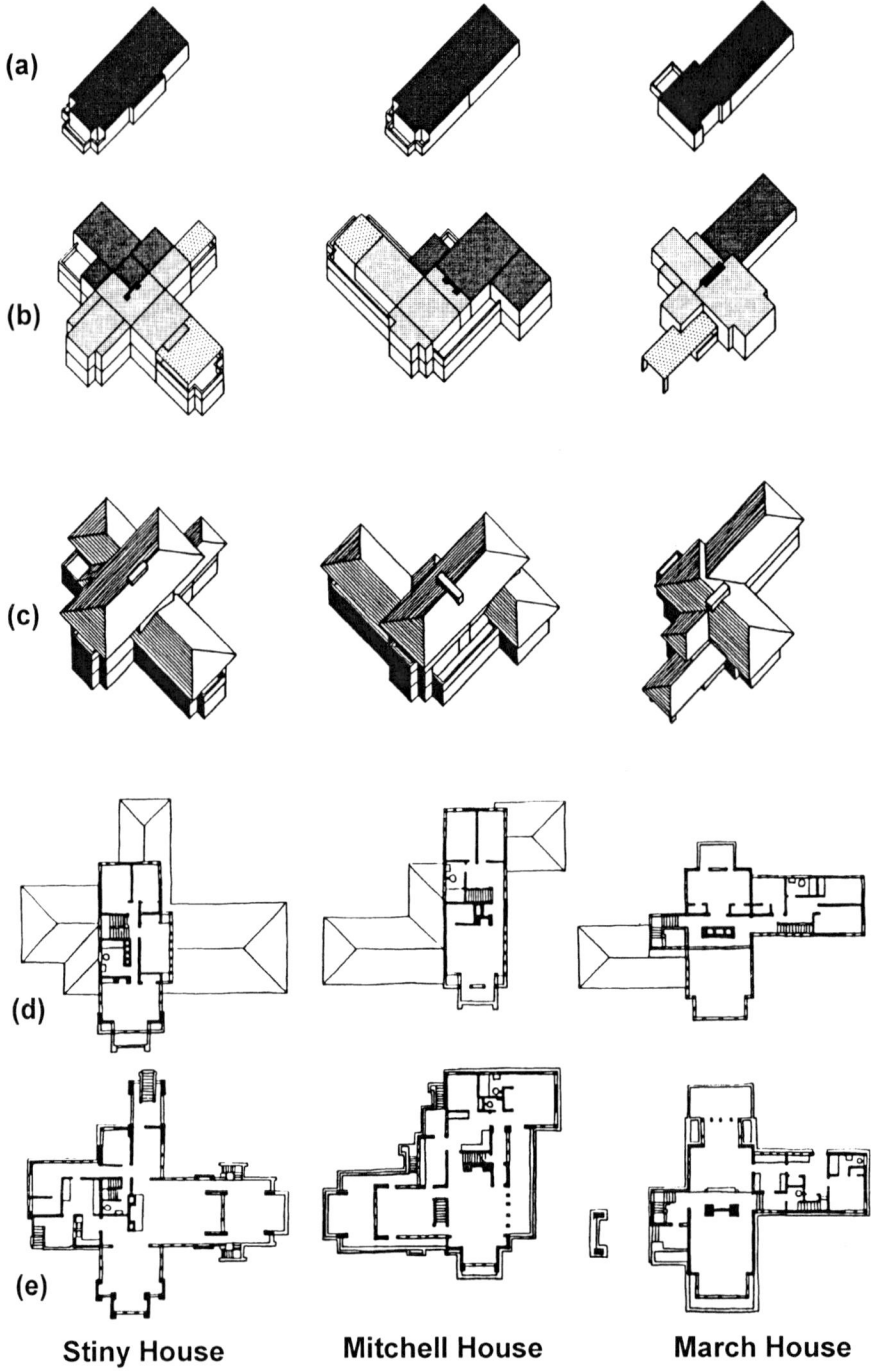

Figure 1.33. Three original designs generated by Koning and Eizenberg's shape grammar: (a) bedroom level, (b) main floor level, (c) external form. Further developed detailed plans: (d) bedroom floor plan, (e) main floor plan.

And here in lies the value and the limitations of shape grammars as design tools. They facilitate the articulation and consistent application of the designer's own grammar of design. Koning and Eizenberg quote Wright himself as follows:

> Consistency in grammar is therefore the property - solely - of a well-developed artist-architect. Without that property of the artist-architect not much can be done about your abode as a work of Art. Grammar is no property for the usual owner or the occupant of the house. But the man who designs the house must, inevitably speak a consistent thought-language in his design. It properly may be and should be a language of his own if appropriate. If he has no language, so no grammar, of his own, he must adopt one; he will speak some language or other whether he so chooses or not. (Wright, 1954, pp. 182-183)

Although shape grammars can be excellent tools for the exploration of a given vocabulary, they cannot of themselves offer a new language.

1.3.3.4 Case-Based Design Aids (CBDA's)

Case-Based Design Aids (CBDA's) have been applied to several design fields including architecture. They provide a means of accessing information derived from previous cases with similar requirements. Archie, developed at Georgia Institute of Technology is an example of a CBDA used to aid in the conceptual design of buildings. It gives the designer access to documentation and evaluations of an inventory of existing design examples relevant to some current work. "The goal is to capture and disseminate lessons learned from design experience, especially when those lessons are not easily incorporated into a field's theoretical framework, and when those lessons bear on the early stages of design." (Kolodner, 1996).

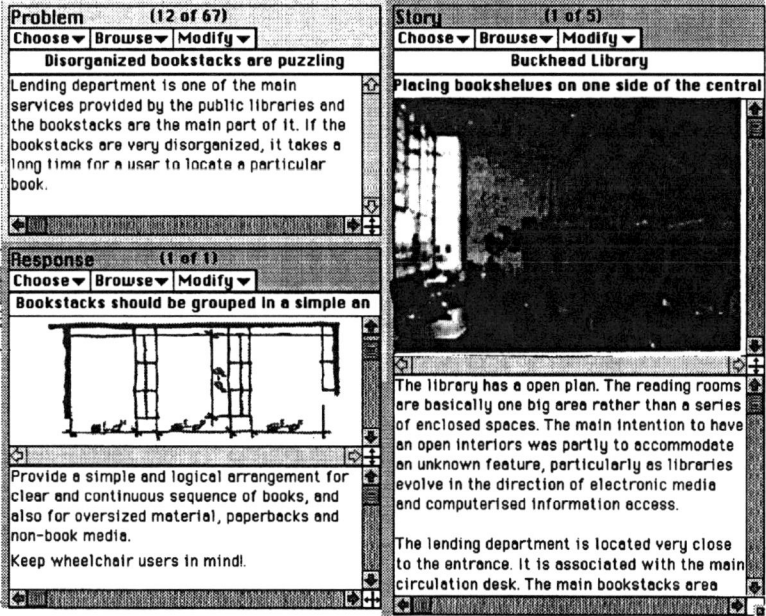

Figure 1.35. An Example Screen from Archie Showing Problem, Story and Response Windows. (Gross, 1994)

The intent is that by making knowledge about previous efforts with similar requirements easily accessible to designers at the early stages of design, fewer mistakes will be repeated and innovative efforts of earlier designers will be brought forward. While this intent may well be achieved, it is not clear that this type of tool really furthers creative, innovative solutions. As discussed in Section 1.2.3.2., the danger of an early infusion of knowledge into the design process is that creativity drops.

1.3.3.5 Intelligent Paper

Mark Gross has coined the term "Intelligent Paper" to describe the use of glyph recognition of freehand sketches to access collections of similar sketches, and provide associative links to a variety of other form defined data bases. In the application, Cocktail Napkin, Gross uses the freehand sketches made by the designer on a special digitizing tablet to access such data bases as: The Great Buildings Collection (a CD-ROM of famous architecture); Archie II (described above); data bases of botanical forms; and the designer's own library of sketches. Cocktail Napkin uses a ranking system based on element types (recognized glyphs), their numbers and their spatial relation to each other. Using this ranking system the various form data bases can then be searched for similar forms. The intent is that seeing other similar forms can lead to analogous associations which may provide richer meaning or new considerations to emerge on the part of the designer.

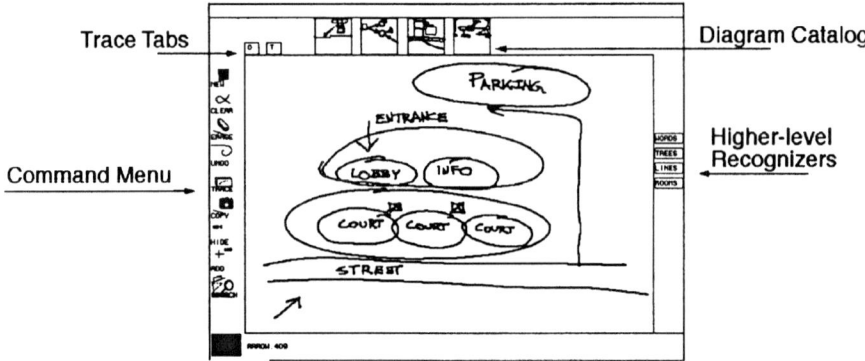

Figure 1.36. An Example Screen from the Cocktail Napkin Program. (Gross, 1994)

Cocktail Napkin itself does not generate any new forms but provides a useful search engine and designer friendly interface to the various data bases. Again there is the danger mentioned above of early infusion of knowledge, and no possibility of encountering truly new forms not contained in the collective data bases.

1.3.3.6 Autogenerated Forms

Autogenerated Forms have been shown to be useful in stimulating designers. Antonio Serrato-Combe has used randomly generated architectonic forms to help his students speculate on possible formal arrangements (Serrato-Combe, 1995). The forms generated, although architectonic in nature, are not specific buildings and do not carry the degree of knowledge with them that is otherwise embedded in a specific design solution. His form generator, instead, is more like a daydream machine. The forms which it generates are abstract enough that they invite the designer to bring further interpretation and meaning to them before they can be seen as solutions. The designer has some control in defining the type of forms generated: lines, arcs, rectangles, 3D

boxes; but the compositions remain random in nature and uncontrollable. Serrato-Combe defines three approaches for implementing his form generators in the design process:
1. The creation of 2-D images generated through random shapes
2. The generation of 3-D images with random geometric solids
3. The simulation of the transformation of complex architectural environments (e.g., housing clusters or townscapes)

Figure 1.37 gives examples of the 2-D and 3-D images. Whereas the method avoids the problem of an over infusion of knowledge in conceptual stages, it remains random and uncontrollable. The program offers no assistance in developing an idea or form further. Each generated form is totally a new trial, and not related to earlier trials. Serrato-Combe's form generator is not intended to develop a specific solution. With the inefficiency of random trial-and-error that could not be expected. What Serrato-Combe hopes to do instead is to stimulate the thinking of a designer sufficiently that the mind will be ripe for a Eureka experience.

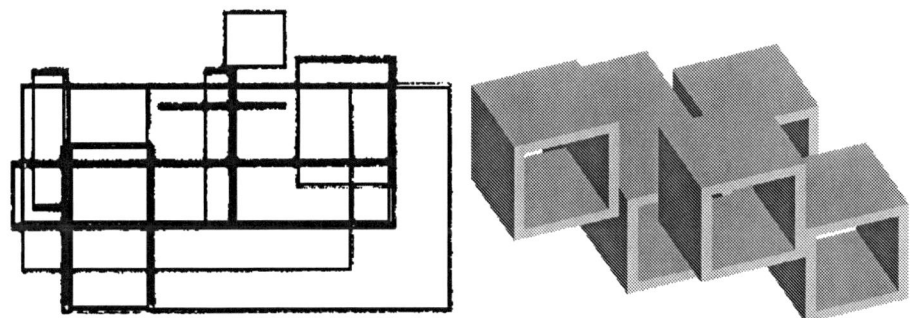

Figure 1.37. Serrato-Combe's 2-D (left) and 3-D (right) Randomly Generated Imagery.

1.3.3.7 Intelligent Evolutionary Systems

Carrying the random generation of form a step further, some researchers have attempted to use evolutionary systems in guiding the generation of form toward some goal. Two examples of this can be found in the work of Celestino Soddu and John Frazer. Another more recent effort is shown by Emanuel Slaby (2003).

Soddu has coined the word Argenìa to describe morphogenetic, generative, evolutionary design (Soddu, 1999, p. 18). In his technique, he makes use of evolutionary algorithms and chaos theory to generate architecturally suggestive forms on the level of individual elements to the level of entire cities. Figure 1.38 shows an example of an environment based on the characteristics of an Italian medieval town, that was generated by Soddu's program.

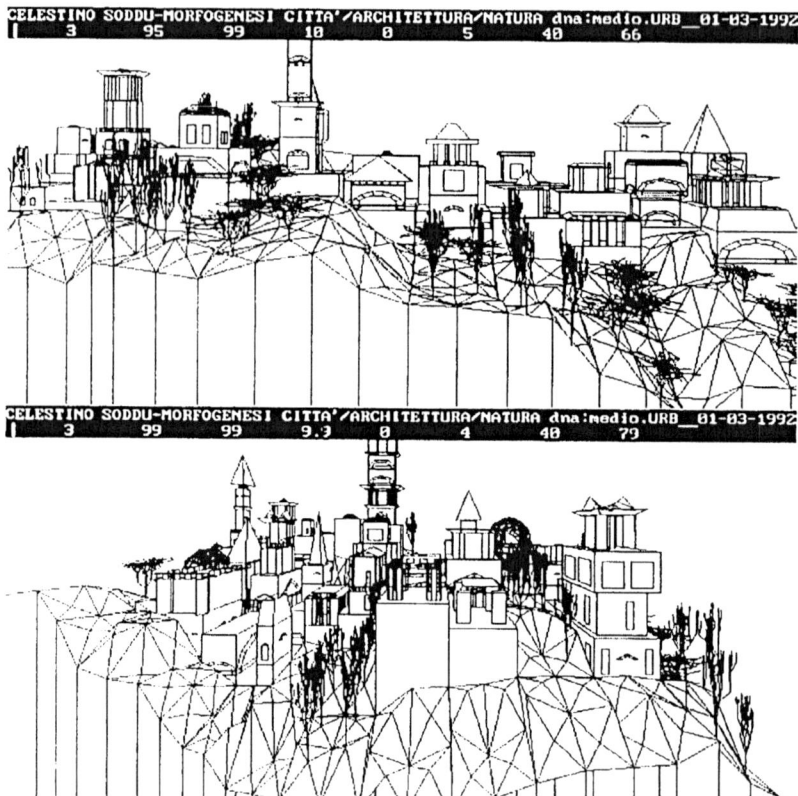

Figure 1.38. Scenarios of a medieval Italian town, generated by Soddu's evolutionary program. (Soddu, 1999, p.21)

> We have designed, as a natural species, the DNA of a typical Italian town, the medieval town. We have referred , for this experimentation, to the drawings of Giotto; and we have interpreted these drawings as an evolving idea of a town environment. To obtain an acceptable complexity of the urban image we have simulated and ran the linear and not-linear dynamics of the evolution of this type of urban image. We have proceeded in identifying and discovering again the rules of the game, the models of the transition between one order to another, the role of randomness in increasing complexity and the power of the time in shaping the environmental image. (Soddu, 1999, p. 20)

In using the software, the designer can adjust a series of variables which then guide the generation of solutions. In this way, the program aids in the exploration of a defined design space by generating and displaying a series of designs found in that space. By changing the initial parameters the designer can alter the design space and shift exploration into other directions.

Soddu has produced a series of similar programs, each encoding the "DNA" of a specific design type, ranging from furniture to architectural buildings, complete with interior spaces. Most interesting perhaps, are results of the actual application of Soddu's design aid. Figure 1.39 shows Soddu's entry in the architectural competition for an enlargement to the Prado Museum in Madrid. Using the form generative program, Soddu found a design solution which he himself claims could not have been realized in any other way. For Soddu, the programming of the tool becomes the artistic expression. (Soddu, lecture at the Symposium on Creative Evolutionary Systems, Edinburgh, 1999)

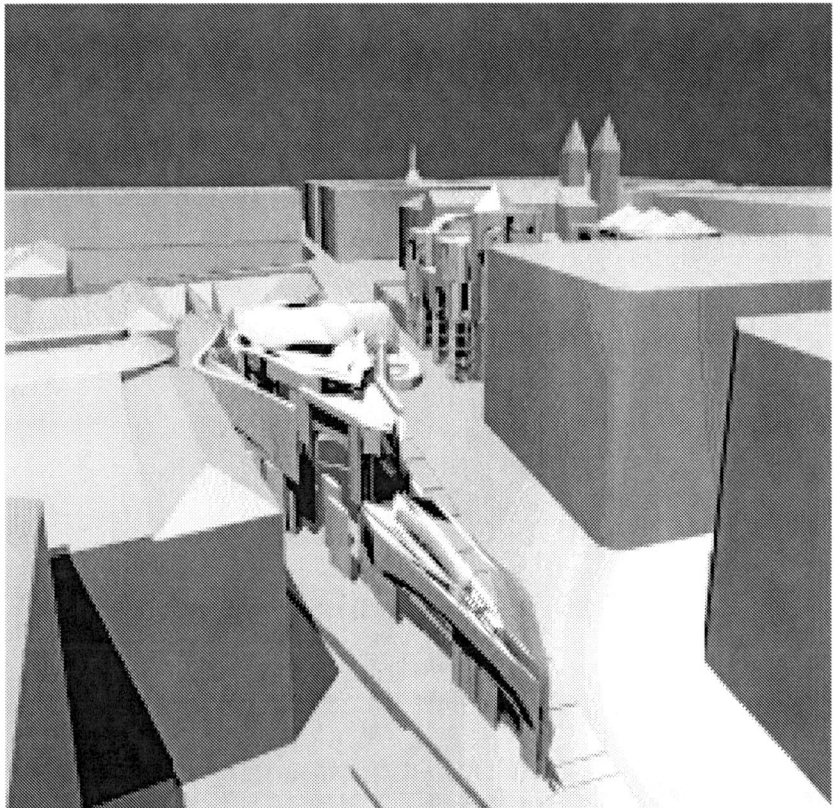

Figure 1.39. Design solution for the enlargement of the Prado Museum, realized with the aid of Soddu's form generating program.

The second example of evolutionary architectural form is found in the work of John Frazer. Starting with a coded architectural genotype, Frazer attempts to evolve phenotypic structures under the influence of a specific contextual environment.

> In order to achieve the evolutionary model it is necessary to define the following: a genetic code-script, rules for the development of the code, mapping of the code to a virtual model, the nature of the environment for the development of the model and, most importantly, the criteria for selection.
>
> It is further recommended that the concept is process-driven; that is, by for-generating rules which consist not of components, but of processes. It is suggested that the system is hierarchical, with one process driving the next. Similarly, complex forms and technologies should be evolved hierarchically from simple forms and technologies. (Frazer, 1995, p. 65)

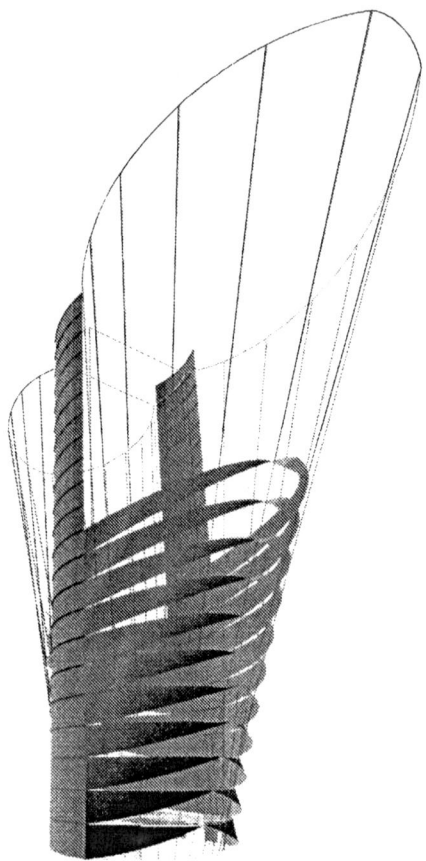

Figure 1.40. "Emergent forms under solar influence: Guy Westbrook, 1993. An initial circular form (which considers all possible sun angles) is allowed to 'grow' with different constraints on its response to the movements of the sun. (Frazer, 1995, p. 64)

In this process the roll of the designer is to specify the original "DNA" of the structure, as described by the five points in the first paragraph above quote from Frazer. Afterwards this "seed" is to be sown in a specific (virtual) environmental context, where it is to grow into a specific instance of a building type. In Frazer's vision, libraries of seed types would be collected and perfected over time, enabling instances of design work to be produced even long after the death of the designer.

Although based on many of the same concepts of genetic design as the IGDT, Frazer's concept is fundamentally non-interactive. Once the initial "seed" is designed, all specific solutions to building types would ideally be generated without further human input. Any creativity required in solving specific instances of a building type would have to be seen as coming out of the program itself. On the other hand, in the concept of an IGDT, human interaction is expected at all levels, and the roll of the program is seen as stimulating the human designer to discover creative solutions.

1.3.4 The IGDT Design Tool

1.3.4.1 Genetic Design Aids

In assisting with conceptual design, genetic methods, and in particular Genetic Algorithms (GA's), do have some potentially useful characteristics that other numerical methods lack.

- Use of Populations
- Recombination and Mutation
- Wide Search of Domain
- No Knowledge of the Fitness Function
- Imitation of Human Design Process
- Adaptable in Nature

Use of Populations. Fundamental to a GA is the use of populations of solutions rather than a single best solution. This is more compatible with the way ideas are often generated and regarded in early design phases. Multiple, simultaneous ideas are necessary for the dynamic movement of thought associated with creative design. Ideas play against each other leading to further new ideas. A designer is more likely to recognize an emergent solution by regarding a group of solutions together. As was stated earlier, creative design is typically not a single-path, linear process, but a more complex, multi-path exploration of many possible ideas.

Recombination and Mutation. A GA uses recombination and mutation to generate new solutions to a problem. This is very similar to design tools developed by William J. J. Gordon (Gordon, 1961) referred to in Section 1.3.1.1. and others, which attempt to combine diverse aspects of different solutions to achieve a more successful design. In a GA, multiple ideas, groups of possible solutions, are collected, compared, cross bred, and combined. One of Gordan's techniques is to fragment and recombine words and phrases. In Synectics one of the "operational mechanisms" suggested to promote creative thinking is to play with words and phrases and their meanings - as a way of making the familiar strange. Gordon quotes Albert Einstein as saying that, "combinatory play seems to be the essential feature in productive thought." (Gordon, 1961, p. 41) This "play" is similar to the mechanisms of random mutation and recombination used in genetic techniques.

Wide Search of Domain. Because a GA contains a certain potential to self mutate, it is constantly searching for different solution directions. It is never completely satisfied in converging on an apparent best solution, but instead is always, to some extent, randomly probing the design domain for other usable solutions. In reference to creative design, George Prince stresses the importance of free speculation as a means of enhancing creative thought.

> We believe that as the expert accumulates the specific knowledge that makes him so valuable he also incorporates the accepted certainties that are not really certain. This explains why, historically, so many innovative breakthroughs have come from outsiders rather than from those who are thought to be most expert in the particular field. (Prince, 1970)

Using mechanisms of random mutation, GA's are able to retain a determinable degree of speculation. In this way they can help even Prince's "expert" to consider new possibilities in the design domain.

No Knowledge of the Fitness Function. Just as biological systems (e.g., plants and animals) are entities separate from their environment, so is a GA separate from its fitness function. Being separate means that the GA is programmatically independent from the fitness function. The GA receives a fitness value (stimulus) from the fitness function, but does not require any information about how the fitness function calculates that value. Analogously, a biological system receives stimulus from the sun, but may have no knowledge of the nuclear fusion in the sun which produces the stimulus. This is a fundamental distinction between GA's and traditional optimization methods based on Linear Programming which require explicit equations describing an objective function. It is this feature of GA's which allows them to be controlled, without pre-programming, by the designer. The GA is steered by means of a fitness function, but the source of the fitness is not necessarily a coded program. The GA can be guided by any means of evaluation, either coded, non-coded or a combination of both. This is very important when considering some hard-to-program, qualitative design criteria such as aesthetics.

Imitation of Human Design Process. A GA learns in a way that is analogous to human learning modes. It seeks a solution by considering many options. Like the near random flow of ideas that a designer will sift through at the start of each project in search of usable concepts, a GA manipulates thousands of solutions, comparing, crossing, recombining, altering, sorting, keeping the best and always scanning the design space for better solutions. The process is not always direct but it is always goal oriented. It is also thinkable that the criteria may evolve as the project matures. As the orienting fitness function changes, the GA automatically adapts and searches in the new direction.

Adaptable in Nature. Like the response of biological systems (e.g., plants or animals) to changes in their environment, a GA will respond to the stimulus produced by the fitness function. Also like biological systems which adapt to changes in their environment, a GA will adapt to changes in the fitness function. As stated above, the fitness function is separate from the GA, and not necessarily coded (see "no knowledge of fitness function" in this Section). Thus, the GA can be guided with criteria that even the designer might find difficult to express. In so far as the designer is consistent in ranking the fitness of solutions, the GA will adapt to the designer's preferences. By considering the solutions proposed by the GA, the designer's preferences may undergo a development of their own. The GA is also able to adapt to these changes in the designer's understanding without direct reprogramming.

1.3.4.2 Definition of the IGDT Concept

An Intelligent Genetic Design Tool (IGDT) is a tool that is able to dynamically adapt to evolving design criteria, through interaction with the designer, to aid in the exploration of a range of good solutions. It is intended to assist the designer in the early, conceptual design phases by anticipating the designer's preferences, and generating a multiplicity of solutions which aid the designer in exploring the design solution space. Although an IGDT uses optimization, it is distinguished from traditional optimization methods in its ability to adapt to non-programmed fitness criteria learned from user interaction. An IGDT is able to adapt using genetic operations based on population, recombination, mutation and selection. Because the objectives of an IGDT are not explicitly pre-programmed, an IGDT can be employed in earlier design phases than is possible with more common analysis tools, without the danger of causing design fixation (see Section 1.2.2.2.) or prematurely restricting the design process. Also, because an IGDT learns from interaction with the user, it is more easily approached by non-computer oriented users. The concept of an IGDT is applicable to many fields involving design. In explaining the concept of an IGDT, this book uses architectural engineering examples focusing specifically on architectural truss design.

Architectural design problems require solutions which consider a wider spectrum of parameters than is ordinarily covered by base functionality and cost. To consider qualitative parameters like aesthetics or meaning, a design tool must remain flexible, and have the ability to adapt to criteria learned from the user. Although in recent years several computer aided design tools have been developed, these tools find application primarily in the later design phases. By offering single "optimized" solutions to the initial parameters used in early conceptual design phases, this type of tool can actually hinder the designer's creative exploration of the design space. The temptation to the designer is to accept the direction offered by the optimization analysis without sufficient exploration of alternatives. Simply having a complete solution presented may lead to design fixation where the presence of one idea tends to block other ideas from being considered (de Bono, 1971; Purcell & Gero, 1996).

The concept of an IGDT is significantly different from traditional analysis and design programs in three ways.

- Not Pre-programmed
- Intelligent
- Exploitative

Not Pre-programmed means that the objective function or fitness criteria are determined or altered as the tool is being used by the designer. In traditional optimization programs, both genetic based and more traditional numerical methods, the fitness, or objective function, is determined in advance, and the solution converges to a solution which optimizes these pre-programmed criteria. In an IGDT the final design criteria are supplied by the user while the program is being used in the form of selection or ranking of individual proposed solutions. These fitness criteria in the form of ranking by the designer, are not pre-programmed. For example the designer may have several qualitative criteria based on concepts of aesthetics, practicality of construction, limits of time, space, skill, and so on. In a real application the designer may have many such criteria that overlap or even conflict. The designer may not even be able to completely verbalize such criteria in words let alone computer code. Nevertheless, in so far as the designer is consistent with the selection or ranking of individual solutions based on this personal criteria, the IGDT will function as a useful tool.

Intelligent means that an IGDT both learns from the designer, and anticipates the designer's direction in the exploration of design spaces. This allows an IGDT to make intelligent, rather than simple random, proposals back to the designer. The criteria used particularly in early design phases are dynamic. An IGDT allows for the adaptation of the design criteria as well as the design results. An IGDT is able to adapt by following implicit criteria learned from the selections made by the designer. It adapts dynamically to the designer's direction, even as that direction may be evolving over the course of a session. This ability to adapt to changing criteria within a session is absolutely necessary for a design tool. It is one of the characteristics which distinguishes design from analysis. By providing the designer with a variety of different solutions, the IGDT stimulates the designer's creativity. On the other hand an analysis tool can only answer questions put to it by the designer. Analysis tools, with the possible exception of Pareto analysis, usually imply a single correct or 'best' solution. (see Section 1.3.2.1.). Since the guiding criteria of an IGDT are learned through interaction with the designer, the IGDT can adapt to these evolving criteria without forcing the designer in a pre-determined direction.

Exploitative means that an IGDT not only searches a single design space for an optimal solution, but by allowing the design criteria to be altered, an IGDT can be used to explore variations of the problem criteria (alternate design spaces) for solutions that better satisfy the goals. Traditional methods with pre-defined objective functions have as a goal the discovery of the single 'optimal' solution. An IGDT seeks *populations* of good solutions with a significant degree of difference. The concept of an IGDT is not only to search for the best solutions defined in one design space, but to explore different possible design spaces in a way that is helpful and stimulating to the designer. It offers both solutions that tend in the direction of the designer's criteria, as well as solutions that explore new directions defined by alterations to the criteria. In this way the IGDT is intelligent in that it is able to offer possibly good solutions in a direction that may not have been previously foreseen. The designer interacting with the IGDT may recognize some such new directions as having potential, and through selection of these solutions allow the IGDT to explore the new design space further. In this way a dialogue exists between the IGDT and the designer in which both share in the exploration of solutions.

1.3.4.3 Relation of an IGDT to the Design Process

The conceptual design process is iterative and usually contains a certain amount of wandering. (see Section 1.1.2.2.) The path toward the final solution is not always direct nor constantly progressive. Tools which are highly directive, which seek a single best solution, can actually hinder design by narrowing the scope of consideration too early in the process, before sufficient possibilities have been explored. People who teach design are very familiar with the tendency students have to latch onto a solution to the extent of refusing to justly consider other possibilities. In the field of psychology this it called fixation or being fixated on a solution. A tool which offers a single 'best' solution runs the danger of causing fixation, and thus setting up a mental block which actually hinders the designer from considering other solutions. For this reason, tools which may be excellent analysis aids, may be very poor design aids.

An IGDT is intended to be compatible with the design process. It allows a certain amount of wandering in exploring different design spaces. It acts as a tool in the hand of the designer. It follows the designer's direction, and responds to the designer's evolving criteria. It is also an intelligent tool in that it is able to autogenerate populations of solutions which are adapted to the designer's own preferences. In this process the designer enters into a dialog with the tool. The IGDT generates reasonable proposals, and the designer critiques the proposals in the form of selections, alterations or further proposals. Through this interaction the IGDT is able to adapt to the designer's preferences, and thus generates further new proposals in the direction of the designer's own interest. In addition the IGDT searches for other reasonable solutions and includes them in the next round of proposals.

This pattern is similar to the way two designers interact in exploring options for a design. There is a certain amount of trading of ideas, as well as a certain amount of individual suggestion. The suggestions from the one may trigger new ideas and suggestions from the other. Well known techniques, such as Osborn's brainstorming (Osborn, 1963) and speech manipulation or Gordon's synectics (Gordon, 1961), are attempts to generate new and creative solutions to a problem. An IGDT fulfills this same roll as an idea generator. The genetic operators of recombination and mutation help to explore design spaces by generating new solutions that respond to the preferences of the designer. Because an IGDT always works with a population of solutions, there is less danger of fixation on a premature solution. An IGDT tends to be expansive rather than constrictive in the design process.

Genetic Algorithms (GA's) are used as the basis for the IGDT because they are conceptually very close in operation to the way many designers work. GA's generate a population of individuals in the way that designers create a pool of ideas from which to draw. GA's operate on this population by recombining parts of different individuals or altering existing individuals though mutation. Designers, too, combine good aspects of various ideas and alter old ideas to fit new situations. GA's explore a multiplicity of design spaces by random sampling, even as the scope narrows toward solutions best meeting the fitness criteria. Designers behave similarly by exploring several options, and by being ever open to new directions for a solution. Also, because a GA has no programmed knowledge of the fitness function, the fitness of the individual solutions can be determined by the designer using qualitative criteria. By not requiring the design criteria to be numerically expressed or directly programmed, the IGDT remains flexible and responsive to changes dictated by the designer.

For most Architectural applications an image of the design is essential to the decision making process (Serrato-Combe, 1995; Gross, 1994). In this book the IGDT is demonstrated using trussed structural systems as examples. The concept presented, however, is not limited to trusses, or even structural systems as is discussed in Section 4.2.1. But, as in any design application, using graphic images in communicating with the designer is important. Work by many authors has shown that activities which involve creative thinking take place primarily in the right hemisphere of the brain (see Section 1.3.1.2.). Processing visual images is also primarily right hemisphere centered. On the other hand, analytic activities, such as language or numeric thinking, are primarily left brain centered. Since it is generally accepted that creative problem solving is more likely to be stimulated by right brain activities, using graphic images to communicate as much of the design information as possible enhances the effectiveness of the IGDT.

In Section 3. examples are give of an IGDT applied to truss design. In these examples the IGDT submits, for the designer's review, images of possible solutions that the user can view, compare, and alter to make new solutions. Being able to view and manipulate the images is essential for the designer's understanding, and allows the designer to make reasonable comparisons and alternate proposals. In addition the images are stored and can be recalled by the designer at any time as a reference, or to resubmit in some form to the IGDT.

1.3.4.4 Outline of an IGDT

Although a complete description of an IGDT is given in Section 2., a brief outline of the overall procedure is given here. The operation of the IGDT is iterative and can continue as long as the designer finds it productive. Like any dialog, it will reach a point of stability were so little new ground is being covered that continuing beyond that point is likely to be fruitless. Because the IGDT interacts 'intelligently' with the designer, later sessions can offer different proposals since the designer's own understanding of the criteria will evolve over time. Therefore, more complex problems may benefit from multiple sessions spaced a few days apart. Inside of one session the activity can be outlined as follows:

- Problem Definition
- IGDT Proposals
- Designer Interaction
- IGDT Response
- Iteration of Previous Two Steps...

Illustrated examples of the entire process are given in Section 2.

Problem Definition. The user initially sets constant design parameters which are used by the IGDT at the start of a session. This step is described in detail in Sections 3.1.1. and 3.1.2. Constant criteria can include material constants and properties; topology constants such as symmetry; geometry constants such as support positions or required load points; required load cases and load combinations; and output specifications.

IGDT Proposals. In the first iteration, the IGDT generates an initial trial set of solutions. These represent either different topologies and/or significantly different geometries. In subsequent iterations, the IGDT develops and proposes new individuals based on the interaction with the designer. The proposals reflect some optimization based on the constant criteria set by the user, but are ultimately guided by the direct interaction with the designer. This step is further described in Section 2.1.3.

Designer Interaction. Presented with a pallet of proposals from the IGDT, the designer begins the iterative dialog which allows the IGDT to adapt the tendencies and preferences of the individual designer-user on the specific project being investigated. The designer precedes by selecting best/worst proposals; modifying some proposals as desired; or creating entirely new proposals. The designer may also alter the initial design criteria from the initial problem definition. This step is further described in Section 2.1.3.

IGDT Response. Using the designer input, the IGDT searches for a new set of proposals using genetic operations of mutation and recombination. Again the proposals represent either different topologies and/or significantly different geometries. But unlike the initial proposals made by the IGDT, all subsequent iterations of proposals are derived from, or influenced by, the interaction with the designer. This step is further described in Section 2.1.3.

As the exchange between IGDT and designer continues, the IGDT adapts to the designer's preferences, whatever those preferences may be based on. The designer is also learning as the range of possible solutions is explored. As new sets of proposals are presented, the images will suggest new considerations and concepts to the designer. In the course of a session, it is expected that the thinking and considerations made by the designer in responding to the IGDT will undergo a development of their own. This is precisely why GA's work so well in the IGDT. Since GA's do not require any knowledge as to how the selection decisions are made, they are free to follow whatever direction is indicated by the designer. As a result the path toward the solution will not be direct. Old solutions may from time to time resurface under changing criteria. This is not important. What is important is that the design space be thoroughly explored and made apparent to the designer. The success of the IGDT depends more on a thorough exploration than on an ultimate 'best' solution.

2 The Intelligent Genetic Design Tool

2.1 Constructing Genetic Tools

The Genetic Algorithms (GAs) used in the building of an Intelligent Genetic Design Tool (IGDT) belong to a class of stochastic numerical methods generally called Evolutionary Computation (EC) or sometimes also called Evolutionary Algorithms (EAs). EC paradigms have been developed by different researchers starting in the late 1950's and early 1960's (Mitchell, 1996. p. 2). With regards to the IGDT it is useful to recognize three categories of EC. The IGDT draws from each of these three groups in its structure.

- Genetic Algorithms (GAs)
    - Genetic Programming (GP)
    - Classifier Systems (CFSs)
- Evolution Strategies (ESs)
    - Evolutionary Programming (EP)
- Interactive Evolutionary Computation (IEC)

Genetic Programming and Classifier Systems are actually subdivisions of GAs. Also, Evolution Strategies and Evolutionary Programming are very similar. All EC paradigms draw in some way upon an analogy to evolutionary genetics with distinctions between groups having in some instances more to do with the history of their isolated development than any major conceptual differences. For an overview of methodology and techniques used in EC, as well as a survey of analogous processes in biological genetics, the reader is referred to a report prepared for the Institute for Lightweight Structures and Conceptual Design (ILEK) at the University of Stuttgart (v. Bülow, 2007).

2.1.1 Design Objectives

In the survey of design tools presented in Section 1.3, a variety of objectives, as well as tools developed for those objectives, were presented. In the following sections, the objectives specific to the Intelligent Genetic Design Tool (IGDT) are outlined.

2.1.1.1 Quantitative Objectives

For the IGDT to be useful as a means of exploring structural form, it must be able to assess and rank, quantitative parameters of any form it might evolve. These parameters define the fitness function for the GA. In the examples presented in this dissertation, parameters of weight (material efficiency) and complexity of form (geometric efficiency) have been used. Although it is possible to include singularly or collectively any number of parameters, care is recommended, as the consideration of too many parameters tends to obscure the meaning of the resulting solution.

It is, of course, possible for the IGDT to run using solely predetermined parameters to guide it in exploring the design space. Even in this mode, the IGDT is more exploratory as a design tool than traditional optimization methods, because it continually presents populations which represent a range of good solutions rather than one instance of an 'optimal' solution. In addition, the 'auto-pilot' mode makes it possible to sometimes reach deeper into the solution space in a shorter amount of time. For example, the program can be set to run unattended through a few hundred topology generations allowing the designer to view a larger variety of solutions. In the graphic output from the automatic mode can be filtered to remove any duplicate solutions and show only solutions that pass a given fitness level (the better solutions). In this way it is possible to control the amount of output and make it easier to see the dominant patterns. It is then possible to proceed from any point reached in the automatic mode by continuing in the interactive mode.

The goal in the automatic mode is to discover as many good solutions to the predetermined quantitative parameters as possible. In this way the user's creative understanding of the solution space is expanded.

2.1.1.2 Qualitative Objectives

It is a unique feature of the IGDT, that it is able to accommodate qualitative objectives as well as traditional quantitative objectives. Qualitative objectives are parameters that are recognizable by the designer, but do not lend themselves readily to quantifiable definition. These include parameters such as aesthetic value, meaning, analogous form, etc. It is characteristic of qualitative values, that they are readily ranked by comparison within a group of examples, but present difficulties for most people to describe independently. For example, given a set of images, it is not necessarily a difficult task to choose the most aesthetically pleasing one. But, to describe precisely why the choice was made, in a way that defines a consistent rule that can be applied to all future choices, is quite a different task.

Guided by qualitative objectives the user makes interactive choices which provide the ranking of the structures found by the IGDT. In this way the IGDT can be guided by the user's own creative curiosity and instinct in exploring the solution space.

Considering the qualitative objectives together with the quantitative objectives, the IGDT can be considered a multi-objective search tool. Multi-Objective Evolutionary Algorithms (MOEA) have been used in a variety of applications including engineering optimization (Coello Coello, 2004). In recent years several approaches to MOEA architecture have been put forward (Mehr & Azarm, 2003) which are generally distinguished in the treatment of the fitness function or the population composition and selection algorithms. The IGDT can be considered an *implicit* multi-objective search tool when guided by the designer's non-coded selection criteria, but it is not explicitly coded to find Pareto non-dominated solutions (a Pareto set).

2.1.2 Encoding Techniques

The search and exploration engine used in the IGDT has been patterned after the CHC Genetic Algorithm, developed by Larry J. Eshelman and J. D. Schaffer of Philips Laboratories (Eshelman, 1991). CHC represents one of the more recent directions in GA development. It combines a highly disruptive recombination operator, which allows for thorough exploration, with an elitist selection operator for good convergence velocity. CHC is successful with small populations (ca. 50). In addition, it makes use of a breeding filter that allows only the fraction of the parent population which promises more productive pairings to produce children. This makes the process of breeding more efficient. These operators combine to give the CHC good exploration qualities, and a high rate of convergence. In the IGDT, good exploration qualities are desirable to provide the designer with a broad view of the solution space. Also, for the process to function interactively, speed is needed in finding good solutions. Since the CHC offers advantages in thorough exploration and rapid convergence, it was chosen as the basis for the IGDT.

The CHC-GA can be seen as running in two, nested cycles. The inner cycle uses an elitist GA with half uniform crossover. The outer cycle begins with a population of mutated individuals based on the best result of the previous cycle. Figure 2.1. shows a diagram of the CHC outer cycle (labeled cycle). The inner cycle is contained in the "run GA" circle.

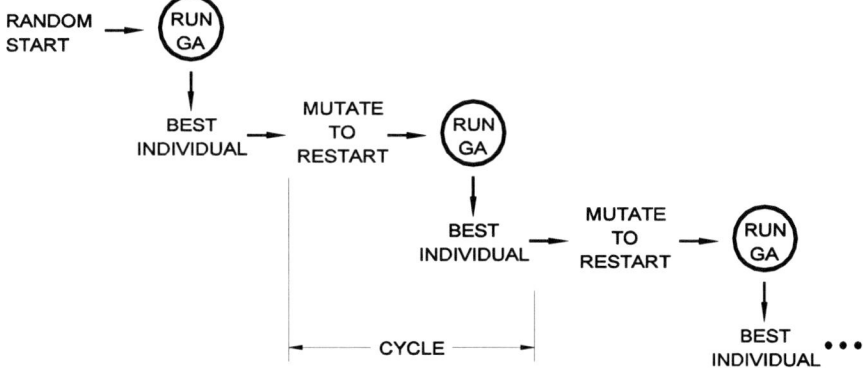

Figure 2.1. The nested cycles of the CHC-GA.

The outer cycle runs a GA a series of times. The very first cycle generally begins with a randomly generated population of parents. The GA is run until the population converges. At the end of each GA run, the best individual is selected, and mutated repeatedly to form a new *restart* population. In contrast to traditional GAs, the CHC does not make use of mutation during the breeding cycle. Instead, mutation is applied only during the restart phase. During restart, the best individual from the converged GA is mutated by randomly flipping a percentage (ca. 35%) of the binary bits in the string. The mutation process is repeated until the new population is filled. One copy of the unmutated source individual is retained in the next cycle as well. The CHC cycle is terminated after a chosen number of restarts are made without improvement to the best individual.

There are advantages in both thoroughness of search as well as computational efficiency which are realized by the cyclic CHC structure. Because the GA used by the CHC incorporates highly disruptive recombination, mutation is ineffective (actually unneeded) in providing diversity to the breeding population. By positioning mutation as a concentrated event at the start of the CHC cycle, new areas of the search space are more likely to be encountered, and exploited by the ensuing GA portion of the cycle.

Also the cyclic CHC structure provides a mechanism for tailoring the search to the degree of difficulty present in a particular problem. In easy problems where the optimum solution is rapidly found in the first few cycles, the CHC will terminate after a few cycles without improvement. In more difficult problems, the cycles will continue to show improvement, and the CHC will continue to explore new areas of the search space by repeated restarts, until a good solution is found. In this way the CHC adjusts the run time to the degree of difficulty of a given problem.

The GA used in the CHC is nontraditional in several aspects. Most significant is the highly disruptive recombination operator, half uniform crossover (HUX), which is used. Uniform crossover, as developed by Syswerda (1989). CHC amplifies the disruptiveness further by providing a filter which allows only parents whose Hamming distance is above a certain threshold to breed. The Hamming distance is a direct

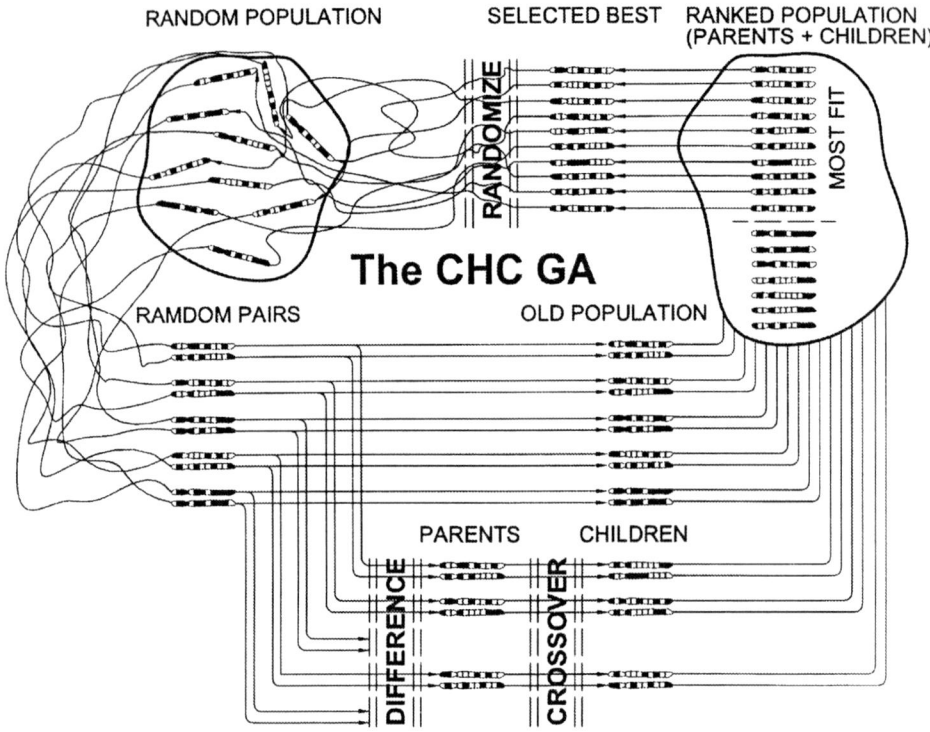

Figure 2.2. Graphic depiction of the CHC-GA.

measurement of the number of differing bit positions of two binary strings. The difference threshold is set so that only the more different (larger Hamming distance) fraction of the parent population is bred. By promoting the recombination of more different individuals, better exploration is insured. By discouraging the recombination of more similar individuals, convergence is retarded by slowing the takeover rate of a group of similar, better performing individuals. At the same time, because only some fraction of the entire parent population is actually breeding, the amount of calculation needed for the generation, analysis and sorting of the resulting smaller child population, is less. Eshelman refers to this filtering mechanism as "avoiding incest" (Eshelman, 1991, p. 273). The difference threshold is initially set at one forth of the binary string length (L/4), which is half the expected Hamming distance of two randomly generated strings. As the population begins to converge the difference threshold is decremented each time a generation occurs which produces no children. In this way breeding is maintained only amongst the most differing individuals in the population at any given time.

The recombination operator HUX is a form of uniform crossover in which one half of the differing bits are exchanged between two parents. This insures a high level of disruption. The resulting two children are both added to the new population along with the two parents, and all subsequently undergo an elitist 'survival' selection (Eshelman, 1991, p. 266). This type of selection, although not typical for GAs, is actually the same as that used by $(\mu+\lambda)$-ES (Bäck et al., 1992). During selection, the parent and child generations are combined, and the best individuals are selected to fill the next generation. The process is strictly elitist. In addition one copy of the best individual is always maintained in the restart population.

2.1.2.1 Describing Topology

Topology is typically described in Finite Element Analysis methods by an incidence matrix which records the connectivity of the elements to the nodes. This matrix can be binary in nature: 1's representing a connectivity between nodes and 0's representing no connectivity. As shown in Figure 2.3., the upper triangle is sufficient to completely describe the topology.

The IGDT converts the upper triangle of the square incidence matrix to a vector array. Although, it is in fact not a true binary string, as in contains integers used to tag the elements, it can be treated in the same manner as a binary string where all of the integers are regarded as 1's. In this way, GA style crossover methods can then be used. A more detailed description of topology encoding and breeding in the IGDT is given in Section 2.2.2.

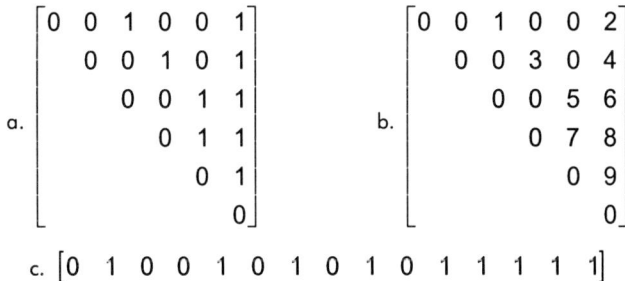

Figure 2.3. The incidence matrix from the example truss of Figure 2.4. a. illustrates the on/off binary nature of the incidence. b. shows how the IGDT uses the same matrix to record a tag to the member. c. shows the vectorized version of the upper triangle for use in GA breeding.

Since different topologies can have different numbers of joints, the size of the incidence matrix, and so the chromosome length, will vary. This is a different situation from the geometry chromosomes which are all the same length. Goldberg proposed an approach to breeding chromosomes with variable length in his "messy Genetic Algorithm", the mGA (Goldberg, 1989b; 1990). The IGDT is similar to the mGA, but does not use over and under specification. One-point cut and splice crossover is used as described in Section 2.1.2.3.

2.1.2.2 Describing the Geometry

Whereas topology describes the pattern of connectivity between nodes, the geometry of a structure includes additional spatial information which describes specific locations of each node. A geometry is thus an instantiation of a topology. Or in genetic terms, geometry is the phenotype of the genotype topology. In the case of trusses, the topology information recorded in the incidence matrix, is supplemented by the vertex matrix, giving node coordinates, to describe a geometry. Further descriptions of member geometries (cross sections or other properties) are contained in additional matrices referenced to the members. In some problems of a more restricted nature, it may be necessary to limit the range of solutions to geometries based on a single topology. This is with the IGDT of course possible, although the more creatively stimulating explorations usually allow different topologies to be regarded as well.

In order to make use of Evolutionary Computing it is necessary to be able to encode the geometry in the form of a 'genetic chromosome'. For a Genetic Algorithm (GA) this record generally takes the form of a binary string. The crossover methods employed by GAs usually rely on this binary coding of the chromosomes. However, due to the spatial nature of the geometry, the breeding methods used typically by Evolutionary Strategies (ESs) seem to offer a more appropriate method. ES also makes use of real numbers directly, rather than binary strings, which also works well with the real numbers which describe the coordinate geometry of the structures being explored. For these reasons the IGDT encodes the geometry in a real number matrix, which is bred using ES techniques.

Specifically in the case of truss structures, the IGDT describes the geometry by means of the vertex matrix, which contains real number pairs (x and y) for each node. Figure 2.4. gives an example of a truss with corresponding vertex matrix.

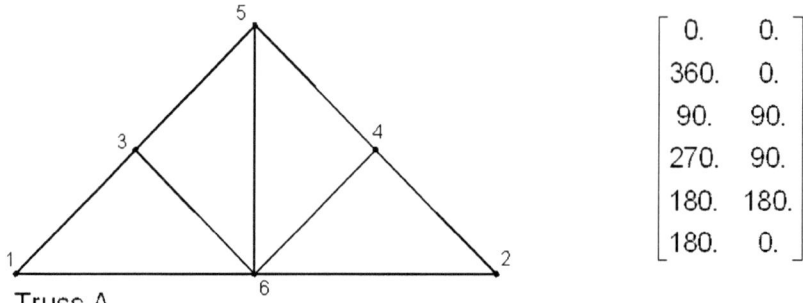

Figure 2.4. An example truss structure with the vertex matrix used to describe the geometry.

2.1.2.3   Crossover for Geometry Breeding

Because the joint coordinates are real numbers with a meaningful spatial location, it is reasonable to incorporate these qualities into the breeding mechanism. During breeding, the joint coordinates of a child are selected from points in a normal distribution about the parent points. The crossover method used in the geometry breeding, is similar to an ES-($\mu+\lambda$) Evolutionary Strategy (Bäck, Hoffmeister & Schwefel, 1992). Figure 2.6. shows how each node is treated as an allele on the chromosome, and half uniform crossover is used for node selection in keeping with Eshelman's suggestions for CHC.

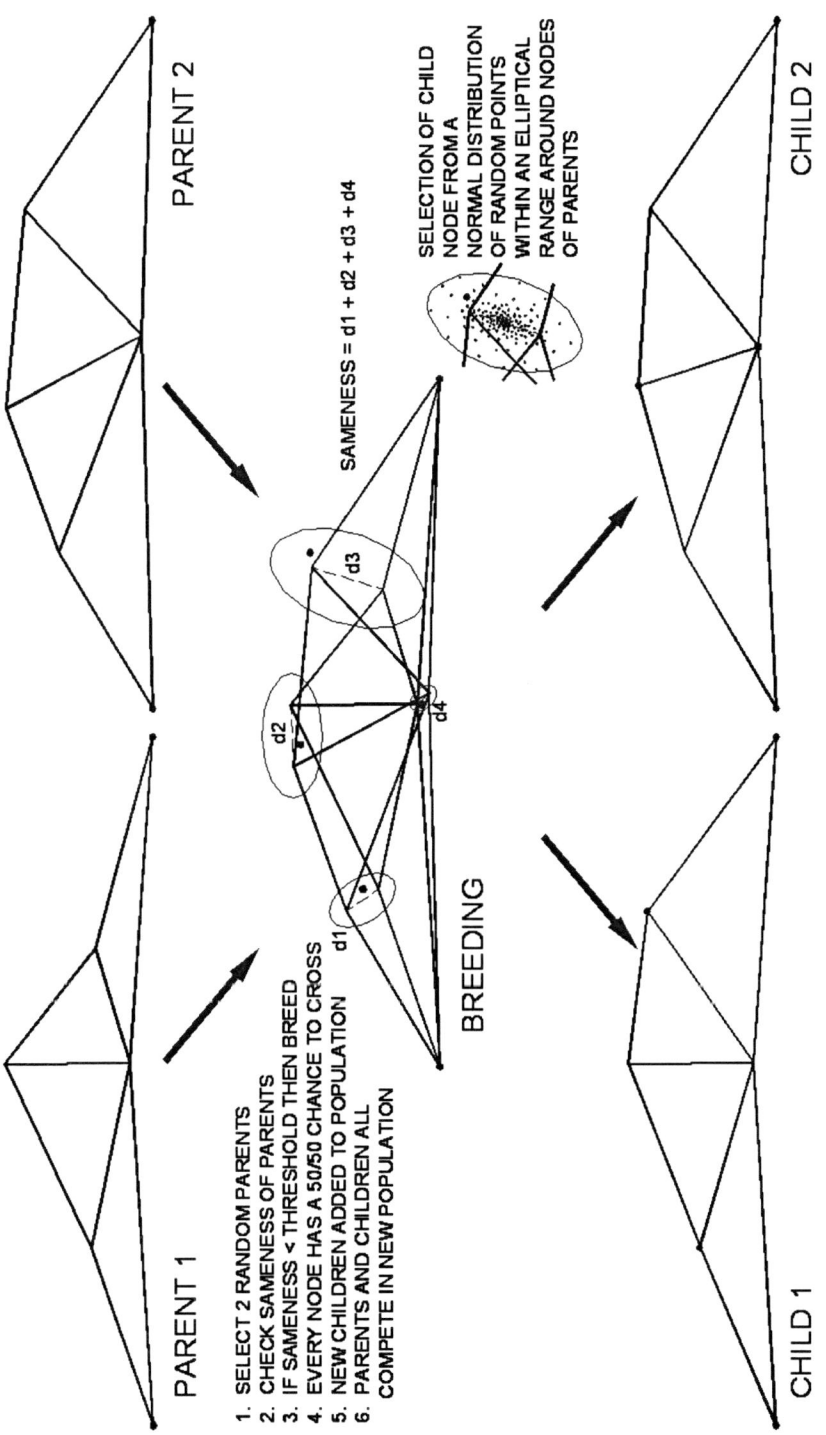

Figure 2.6. The breeding of two parent geometries, and the calculation of a difference threshold.

2.1.2.4 Crossover for Topology Breeding

In the topology breeding mechanism, one-point crossover is used as it is the least disruptive. Less disruption is preferred, because combining the often unequal topology chromosome lengths results in additional disruption itself. In traditional GAs, as in nature, breeding unequal length chromosomes is a problem. In the IGDT the crossing of the unequal length chromosomes is achieved by ensuring that the crossover point is always chosen within the length of the shorter chromosome. Figure 2.5. shows the one-point crossover applied to the vectorized incidence matrix.

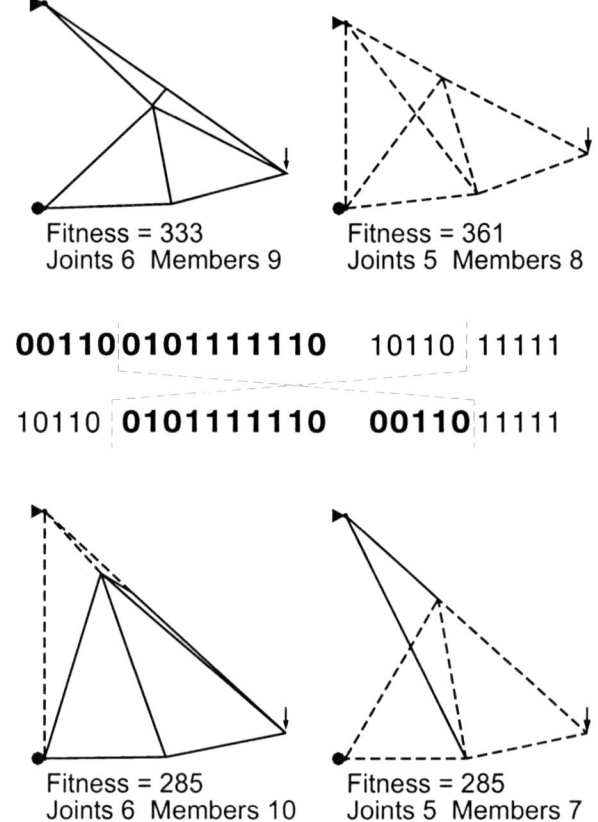

Figure 2.5. The breeding of two parent topologies with different length defining chromosomes.

2.1.3 Search and Exploration

2.1.3.1 Finding Low and High Peaks

Section 1.2.3 describes the stifling effect that viewing a single solution can have on problem exploration, and how this in turn inhibits creative thinking. The IGDT avoids this problem by always presenting a pallet of solutions to the user. These solutions are, however, not merely random geometries, which would at best offer a search guided only by chance, but are instead, a pallet of solutions found by the IGDT, which represent a series of sub-optimal peaks in the solution space. This is particularly the case in the exploration of topology, where each individual in the population has been optimized, and represents a geometry which reflects the quantitative objectives. In the topology

IGDT, each generation offers the designer a set of solutions which can vary greatly. To further enhance variation presented to the user, the topology IGDT can be set to prevent duplicate solutions occurring in the same generation. Duplicate solutions normally occur as the population converges on an optimum. The result of preventing duplication, is that even more variation is presented to the designer. This may provide enhanced exploration qualities, but it also prevents convergence.

2.1.3.2 Repairing Defective Topologies

Nature provides mechanisms for the repair of biological genetic material. If these repair mechanisms were not in place, the instances of mutation detrimental to the organism would be very high (Russell, p. 564, 1992). As a result of initial random generation as well as subsequent breeding of individuals, topologies can be formed which are structurally unstable. It would be possible to simply allow these misfits to be assigned a low rank in the population, and thus be eliminated in the next generation. In fact, this is the procedure followed with unstable geometries which are produced. But, in the case of structurally defective topologies, the IGDT, like nature, has a few repair mechanisms that allow certain defects to be detected and repaired.

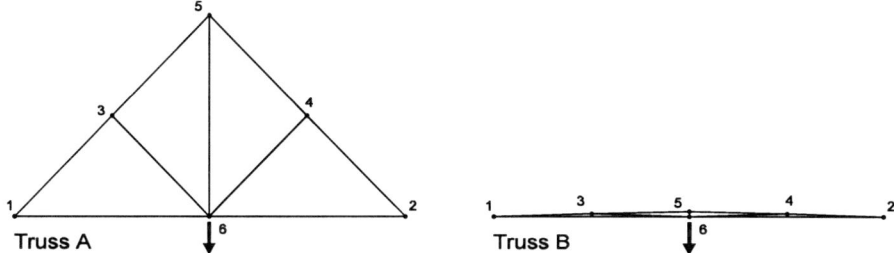

Figure 2.7. Two trusses each with the same topology. Truss A shows a stable geometry while Truss B shows a geometry that is not stable.

For any given topology, geometries can be found that under load produce deformations of such large magnitude that the system becomes unstable. Figure 2.7. shows an instance of a topology first with a stable geometry, and second with a geometry that is not stable. In these cases the unstable geometries simply die out in the selection of the next generation, and the exploration of the topology continues with the remaining stable solutions.

However, in the case of topologies that contain an inherent flaw, all of the geometries in the population will be unstable. Figures 2.8. and 2.9. show two examples of defective topologies. There are three reasons why it is better to repair these defective topologies if possible. Firstly, a population of geometries based on a defective topology will not converge. This can lead to the waste of a large amount of computing time and needlessly slow the IGDT's progress. Secondly, because the topology populations tend to be smaller in size, particularly if the interactive mode is being used, dead individuals are more critical, because they remove a larger percentage of the population from useful exploration. Thirdly, through the repair mechanism, a new topology can be introduced into the population that might not be attainable through breeding alone. This action is comparable to a beneficial mutation that allows new genetic material to enter a population.

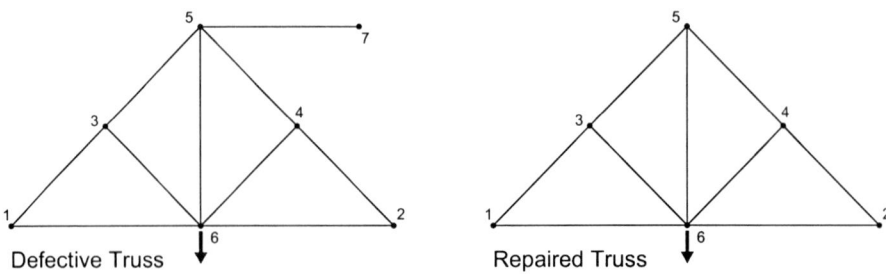

Figure 2.8. An example of a truss with defective topology. Node 7 is not stable and is deleted from the truss.

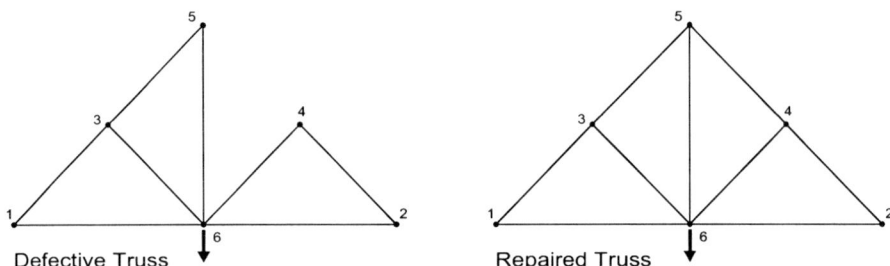

Figure 2.9. An example of a truss with defective topology. Element 5-4 is added for stability.

The two repair routines used in the IGDT entail the deletion of ill-connected joints and the addition of members which are lacking. Figures 2.8. and 2.9. show respectively the results of the application of these two routines to the defective topologies.

2.1.3.3 Adapting New Topologies

Adaptation describes change in response to the environment. For the IGDT the environment is described by both the qualitative and quantitative objectives which guide the search. In response to the objectives, a structure may evolve one topology so that it begins to resemble another topology. In the truss exploration problems, the IGDT uses two forms of adaptation, one that fuses nodes , and a second that fuses elements. Figure 2.10. shows an example of node fusing. Nodes 3 and 7 in Truss A are so close that the topology of Truss A looks like the topology of Truss B. In the figure the distance between the nodes 3 and 7 is somewhat exaggerated to make the duplicity clear. Node fusing is performed by combining the two nodes into one node, which is shown as node 3 in Truss B. Also note that the four elements in Truss A, which connect nodes 1, 3, 5, and 7, are reduced to two elements in Truss B, as node 7 is removed, leaving no two members with the same incidence.

Adaptation provides an important enhancement to exploration. By deleting nodes, it allows new length chromosomes to enter the population that would not occur through breeding alone. In this way much grater areas of the solution space can be reached and explored.

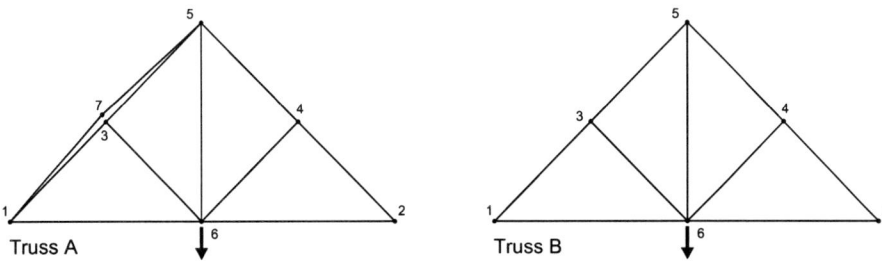

Figure 2.10. A truss shown before and after adaptation by node fusing.

The second type of adaptation used is that of element fusing. Like node fusing the goal is to remove redundancies in the topology. Elements which overlap and are situated on top of nodes, but without connecting to the nodes, are broken at these existing nodes to form new shorter elements. The resulting elements which have identical incidences are fused to form one element. Figure 2.11. shows an example of this process.

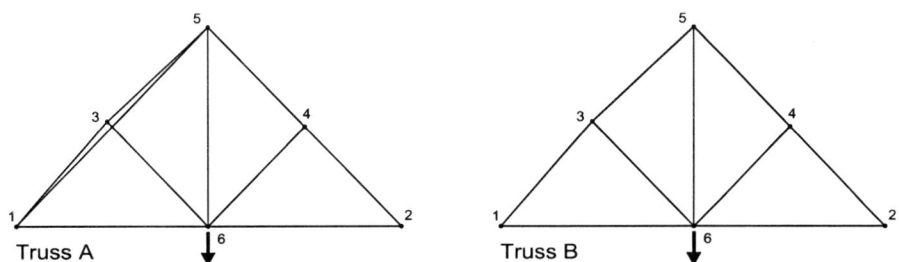

Figure 2.11. A truss shown before and after adaptation by fusing a node into an element.

2.1.3.4 Mutating Topologies

In the CHC-GA, each outer shell cycle begins by repeatedly mutating the best individual from the end of the previous cycle, to produce the initial population for the new cycle. Mutations are simply random changes to the genetic description of the individual, in this case the incidence matrix which defines the topology. There are eleven different mutation operations used in the IGDT. Each operation is applied to the topology incidence matrix. Several also use information from the specific geometry of the individual being mutated. Distinguishing the mutations by whether they make use of geometry information or solely topology information, they can be listed as follows:

Geometry Based Mutation	Direct Topology Mutation
mirror left side	add joint
mirror right side	delete joint
fuse nearest joints	add member
fuse nearest members	delete member
	move member
	flip row
	flip column

The geometry based mutations do not make use of random selections and are therefore only performed once in generating a population. The direct topology mutations are based on random selections and are repeated in random order until the population is filled. In addition to these 11 basic mutations, many more are actually available by combining operators. For example, a combination operator might be: add joint + mirror left side.

The mirror left side operator uses a vertical center-line to cut the selected individual into two sides, left and right. The right side is then deleted and replaced with the mirror image of the left side. If a joint is within some tolerance to the center-line, it is repositioned directly on the center-line and not mirrored. Members that cross the centerline and thereby loose the right end joint, are redirected to the mirror of the left end joint. Figure 2.12. shows the results of the mirror left side operator.

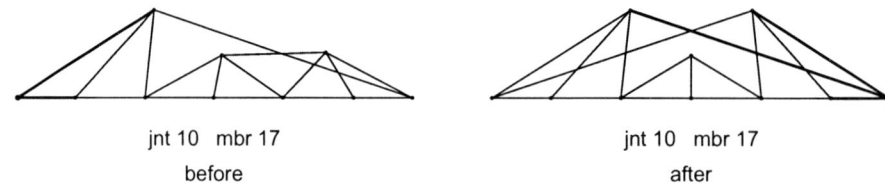

Figure 2.12. Example showing the application of the mirror left operator.

The mirror right side operator is the reverse of the mirror left side operator. It also uses a vertical center-line to cut the selected individual into two sides. In this case, the left side is deleted, and replaced with the mirror image of the right side. Joints and members near the center-line are treated as with the mirror left side operator. Figure 2.13. shows the results of the mirror right side operator.

Figure 2.13. Example showing the application of the mirror right operator.

The fuse nearest joints operator sorts though the joint coordinates and finds the two joints with the closest proximity to each other. It then merges these two joints into one and reconnects all members that connected to the original two joints to the new single joint. Figure 2.14. shows the results of the fuse nearest joints operator.

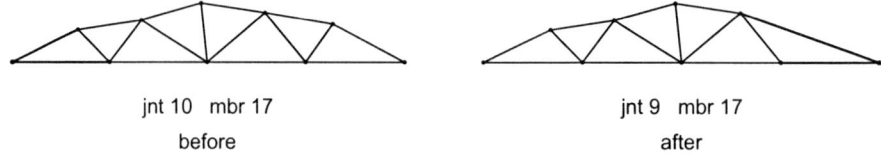

Figure 2.14. Example showing the application of the fuse nearest joints operator.

The fuse nearest members operator sorts though the joint coordinates and finds one with the closest perpendicular distance to some member. It then divides the found member in two and fuses the new joint and the original joint. All coincident members are reduced to single members so that there are no duplicate members. Figure 2.15. shows the results of the fuse nearest members operator.

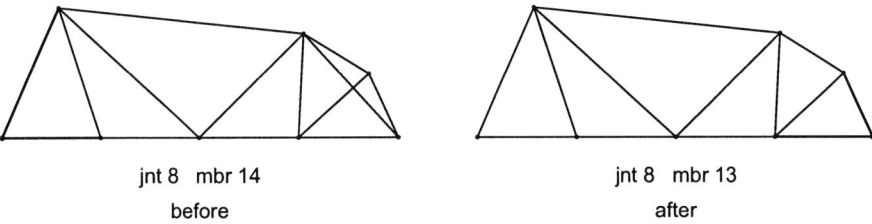

jnt 8   mbr 14   jnt 8   mbr 13
before   after

Figure 2.15. Example showing the application of the fuse nearest members operator.

The add joint operator randomly selects a member and adds a joint at the mid-point. It then finds the closest joint to the newly added joint that is not on the same line, and connects these two joints with a new member. This is necessary to preserve stability. Figure 2.16. shows the results of the add joint operator.

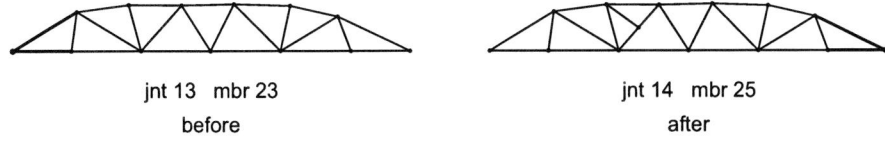

jnt 13   mbr 23   jnt 14   mbr 25
before   after

Figure 2.16. Example showing the application of the add joint operator.

The delete joint operator chooses a joint at random and removes it. The free end of the members which had been attached to the now missing joint, are connected to the next nearest joint. Members which thereby have both ends located at the same joint (zero length) are removed. The resulting truss is checked for stability and repaired if necessary (see Section 2.1.3.2). If the repair results in a return to the original topology, the operation is taken as failed and the results discarded. Figure 2.17. shows the results of the delete joint operator.

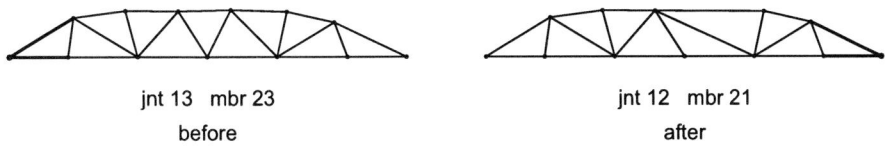

jnt 13   mbr 23   jnt 12   mbr 21
before   after

Figure 2.17. Example showing the application of the delete joint operator.

The add member operator selects two joints at random and tests to see fit a member already connects the two joints. If no member is found between the joints, a new member is added, otherwise two new random joints are selected. If no new member can be found after a number of attempts equal to twice the number of joints, then the operation is taken as failed and the results discarded. Figure 2.18. shows the results of the add member operator.

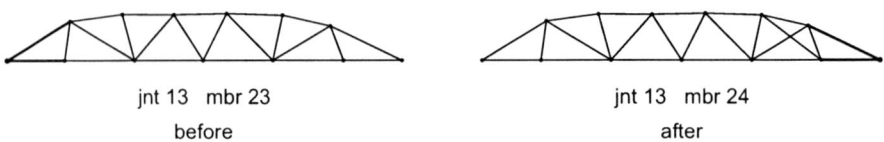

jnt 13 mbr 23   jnt 13 mbr 24
before   after

Figure 2.18. Example showing the application of the add member operator.

The delete member operator selects a member at random and removes it. The resulting topology is checked for stability and repaired if necessary (see Section 2.1.3.2). If the repair results in a return to the original topology, the operation is taken as failed and the results discarded. Figure 2.19. shows the results of the delete member operator.

jnt 10 mbr 17   jnt 9 mbr 15
before   after

Figure 2.19. Example showing the application of the delete member operator.

The move member operator selects a member at random and moves one end to another randomly selected joint. The resulting topology is checked for stability and repaired if necessary (see Section 2.1.3.2). If the repair results in a return to the original topology, the operation is taken as failed and the results discarded. Figure 2.20. shows the results of the move member operator.

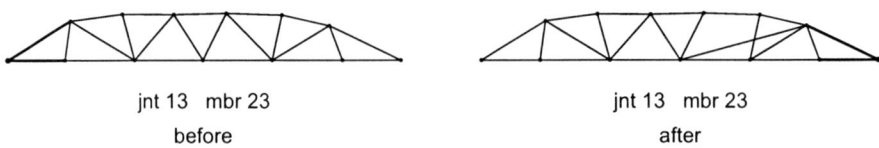

jnt 13 mbr 23   jnt 13 mbr 23
before   after

Figure 2.20. Example showing the application of the move member operator.

The flip row operator reverses the order of one row in the incidence matrix which describes the topology of the truss. The resulting topology is checked for stability and repaired if necessary (see Section 2.1.3.2). If the repair results in a return to the original topology, the operation is taken as failed and the results discarded. Figure 2.21. shows the results of the flip row operator.

Figure 2.21. Example showing the application of the flip row operator.

The flip column operator reverses one column in the incidence matrix. The resulting topology is checked for stability and repaired if necessary (see Section 2.1.3.2). If the repair results in a return to the original topology, the operation is taken as failed and the results discarded. Figure 2.22. shows the results of the flip column operator.

Figure 2.22. Example showing the application of the flip column operator.

2.1.3.5 Storing Results

Because it is expected that the IGDT would be used over a longer period to explore families of structural geometries, a history of significant geometries is maintained in the form of an output file which contains all pertinent geometric and structural data. The data may be viewed in different ways by the designer as a form of review, either as a graphic CAD file or detailed text listing. Also, stored solutions may be retrieved, and inserted into a running population during the interactive mode.

2.1.3.6 Communicating with the Designer

Key to the success of the IGDT is the way in which the program interfaces with the designer. The IGDT is intended to provide the designer with a window into a landscape of structural forms. As discussed in Section 1.2, in order for this window to foster the designers own creativity, certain considerations need to be observed in the formulation of the user interface. Primary considerations include:

- represent a pallet of solutions
- allow alterations by designer
- render with simple depictions

Pallets of Solutions. As discussed in Section 1.2.3.3, design fixation is a common and serious impediment to creative design. Particularly in dealing with computer generated images, the tendency is to show the user only one solution at a time. It might be argued that this stems from the analysis technique itself, which strives to find the best solution to the objectives which have been chosen. But this does not wholly account for the situation. Certainly, it would be possible to show solutions within a range of the objectives. But the true problem seems to stem form the fact that most programs which designers attempt to use, have been conceived as analysis tools rather than design tools. Simply by always presenting the designer with a pallet of solutions, rather than one single solution, the tendency for design fixation is greatly lessened. By allowing the designer to observe an array of solutions, the eye naturally looks for similar patterns among the individuals, rather than concentrating on one single image alone. In this way the user is invited into further exploration of alternatives rather than becoming fixated with a single solution.

Alterations by Designer It is also important to provide a mechanism for the designer to be able to directly alter the forms which the tool generates. The designer should not remain simply a passive observer, but should interact directly with the form generation by altering given forms or inserting new solutions into the breeding population. The IGDT accommodates this interaction in several ways:

- select favorite solutions for breeding
- explore a single solutions through mutation
- alter a current solution
- enter a new solution
- recall and old solution

User selection, either for breeding or mutation, is the primary way for the IGDT to be guided in the interactive mode. The user can select any of the individuals presented from the combined parent and child populations, for breeding in the next generation. Parents can be duplicated or even paired with new user-inserted individuals. If further variations in the direction of one single individual are desired, the mutation option will fill the combined parent and child populations with mutations of the chosen solution.

Being stimulated by the solutions generated by the IGDT, the user will often want to explore specific alteration to a current solution. The IGDT gives the user immediate feedback by evaluating the altered solution based on the same quantifiable objectives defined for the problem. It is also possible to run the geometric optimization on the altered topology. In the same way that alterations to a solution are interactively submitted, an entirely new solution can be injected into the breeding population by the user. Finally, since a history of each topology solution is written to a file, an old solution may be recalled from this file at any time. This includes possible collections of solutions from earlier runs.

User intervention of this sort may seem foreign to many seasoned optimization programmers, but one must realize that it is the allowance of this more playful treatment of the design procedure that gives the IGDT an advantage in attaining more creative solutions through encouraging a more thorough exploration of the design space. Once the designer has gained a sense of the possible solutions using the IGDT, it may well be appropriate to continue the design process by exploring some of the discovered solution further with more traditional optimization techniques.

Simple Depictions Finally, it is intentional that the depiction of the solutions remain as simple as possible. In the truss examples color coding is used to indicate whether an element has been designed for tension or compression, and the material expended is indicated by the line weight. However, excessive detail that would only distract from the intent, is omitted.

2.2 Implementation of the IGDT

The following section gives a more systematic and detailed description of the actual programmed implementation of the IGDT. Conceptually, it can be applied to a wide range of structural types, materials and loadings. However, to illustrate the concept, this implementation focuses on two dimensional (flat plane) truss structures.

2.2.1 Defining Problem Parameters

Parameters specific to a particular problem are entered in an open format text file typical of traditional finite element program input data files. Examples of input data files are included in Appendix A.

2.2.1.1 Structural Type

The structural type used in all examples in this dissertation is a two dimensional flat truss. As such, all elements are simple, straight, axial force members, acting either in tension or compression. All nodes are considered pinned. The finite element used for the stiffness analysis has two degrees of freedom and is commonly found in most introductory text books to finite element analysis, FEA (Martin, pp. 28-54, 1966). Structures to be analyzed may be either determinate or indeterminate. Checks are made during the analysis of each structure to ensure a non-singular (stable) stiffness matrix.

2.2.1.2 Material

For the truss analysis, four material properties are needed:

- Young's Modulus of Elasticity
- Cross Sectional Area (initial estimate)
- Elastic Limit
- Density

Young's modulus and the cross sectional area are needed for the stiffness analysis. If member self weight is not to be included, then the estimate area supplied is used in the FEA. The area of each member is updated after an initial calculation of member forces. If member self load is being considered, the FEA is iterated, and member area is updated until the area stabilizes within 2%.

The elastic limit is used in calculating the member cross section when buckling analysis is to be included as explained in Section 2.2.1.5 below. Density is used to calculate the weight for self load and assessment purposes.

2.2.1.3 Supports

There are two types of node restraint. The first is a support in the traditional sense of an FEA. For a truss type structure this means fixity in either the x or the y direction. In addition, nodes can be prevented from being relocate during various genetic manipulations which can take place. In other words, nodes can be held stationary in either x or y or both directions. This is essential when only one component of a node is supported, but it is expected that the node will remain at a specified location. Another instance which requires a stationary node is the location of point loads. It must be remembered that in the IGDT the structural geometry and even topology will change, and therefore, if some locations need to be maintained, they will have to be specified. Figure 2.23. shows instances of supported nodes and stationary nodes which make the use and distinction clear.

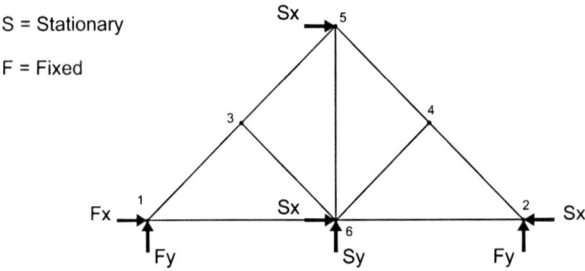

Figure 2.23. Example showing the use of traditional structural supports and geometric restraints.

2.2.1.4 Loads

Loads are of only two types, point and self load. However, loads can be grouped in 'Loadings' (e.g., snow load, dead load, live load) and then applied as combinations of loadings (e.g., DL+SL, DL+LL, etc.). In this way it is possible to design the system regarding the worst load case for each member. For example, a bridge design can be regarded under a series of point loads that move across the structure (rolling truck load). Each member in the structure will be designed for the loading that produces the most severe conditions for that member.

2.2.1.5 Analysis

The analysis proceeds in two steps. To obtain member forces and displacements, a finite element stiffness analysis (FEA) is performed. The element type would be set as described in Section 2.2.1.1 above. The second step of the analysis determines the member cross section given the force in the member and the member length. There are several routines that can be chosen for the member design, based on material, whether the member is in tension or compression and whether buckling of compression members is to be considered. Cross sections can be selected from a table (which is somewhat slower) or estimated by an equation. At present routines are written for steel and wood design by US codes: The Manual of Steel Construction - ASD (1989) and The National Design Specification for Wood Construction (1992). There is also a simple area=force/stress routine for comparison with some published optimization results.

2.2.2 Topology Search

As described in Section 2.1, the IGDT is comprised of two components, an outer cycle that explores topology, and inner cycles that explore geometry. The program allows a useful degree of control at each level in order to balance the thoroughness of the exploration against necessary time constraints (the duration of a run).

The topology parameters are applicable to all geometries generated in the course of a run. These parameters include the problem parameters discussed in Section 2.2.1, which are contained in the input data file, as well as #define statements set in the header file, truss.h. The topology parameters defined in the truss.h file are discussed in this section.

2.2.2.1 Program Mode

The IGDT can run in several modes. The run modes include:
- interactive with user
- automatic with preset number of cycles
- automatic with convergence goal

Interactive with user mode is described in Section 2.1.3.5. This mode will continue to cycle through topology generations until the user chooses to terminate. At the end of each cycle the user can choose to:
- select pairs of parent topologies to breed
- select one parent topology to mutate
- insert a new or altered topology into the population
- quit the program

Automatic with preset number of cycles mode allows the user to set the IGDT to explore several generations of topology unattended. This may be useful as an initial way to view a larger sampling of the solution space. A file is maintained of each different topology and significantly different geometry found. As the run proceeds the file is updated with the best instances of each solution type. This grouping can be viewed graphically as well as in a more detailed text output for each solution in the group. In this way a large quantity of solutions, which are mostly repetitive, can be distilled into a digestible number of unique solutions to consider.

Automatic with convergence goal mode is used for an even longer search of the design space. A convergence criteria is set by specifying a percent of the population that must converge before the next cycle begins. After a set number of cycles converge on the same solution, the run is stopped. This is similar to the way the geometry CHC-GA runs. The same output file described above are also generated.

2.2.2.2 Population Size

The required size of the topology population varies with the complexity of the problem and the mode chosen. For a typical GA, recommendations vary from 50 to 500 (Mitchell, 1996, p. 11). But populations of this size would be difficult for a user to reasonably handle in making visual selections when using the interactive mode. Also, since each topology is in turn explored by a GA, even a small topology population will require a large amount of computing time. Therefore, it is necessary to be able to set the population size to fit the available computing resources. This is possible with the #define TOPO_NR in the header file.

2.2.2.3 Topology Limiting Parameters

In order to insure reasonable limits to the size of the topologies generated, some limits need to be set. These too have been placed in the header file. There are currently two parameters. JOINTS_MAX defines the maximum dimensions of the incidence matrix.

This must of course be larger than the minimum number of joints required for the nodes defined as either fixed or static in the input data file. The maximum and minimum number of members are calculated based on the joint limits.

The second limit defined by HUB_NR, sets the lower limit for the number of elements that attach at a node. For trussed systems this is normally set to 2. The value is used in checking the nodes for stable attachment to the system, discussed in Section 2.1.3.2, and illustrated in Figure 2.8. If the value is set to 1, then only joints completely disconnected form the system will be deleted during the chromosome repair procedure.

2.2.3 Geometry Search

The program parameters are the controls used to vary the performance or output of the program. The program can be used in several different ways, either to explore structural topology or geometry in either an interactive or an automatic mode. Because of this varied use, and because of the developmental nature of the program, it is necessary to have different output possibilities both to the screen and to various files. These output options have been placed in the code using #ifdef statements, so that they can be selected at compilation.

Other parameters which can be used to tune the IGDT by setting certain variables or choosing between optional subprograms, are grouped together in the header file truss.h. These parameters are used to configure the IGDT, and do not need to be changed with every problem. These configuration options are categorized below depending on whether they pertain mainly to the geometry or the topology search.

The geometry GA searches for solutions that perform well when assessed by the predefined fitness function. This GA is based on the CHC-GA developed by Larry Eshelman (Eshelman, 1991). The CHC repeats the cycle of Restart $\rightarrow$ Run GA $\rightarrow$ Select as shown in Figure 2.1.

The #define section of the truss.h file is given in Appendix A and lists all of the parameters that can be set with brief definitions of their use.

2.2.3.1 Restart

A restart is made by mutating the nodes of the last best individual to fill a new population. To form each newly mutated individual each movable component of each node is mutated at a probability of 50%. This is similar to joint selection in Half Uniform Crossover, meaning about half of the node components are altered, while the other half are not altered. The mutation is accomplished by moving the node coordinate some random distance. The possible spread of this mutation is lessened with each restart. There are two choices for the random number distribution, either uniform or normal distribution from the original node. Figure 2.24. shows the two distribution patterns for 200 selections. It is also convenient in some problems to restrict the generated coordinates to only positive numbers. This also is shown in Figure 2.24. There are then a total of four ways in which the node mutations can be configured.

Uniform distribution:
- + and - values
- + values only

Normal Distribution:
- + and - values
- + values only

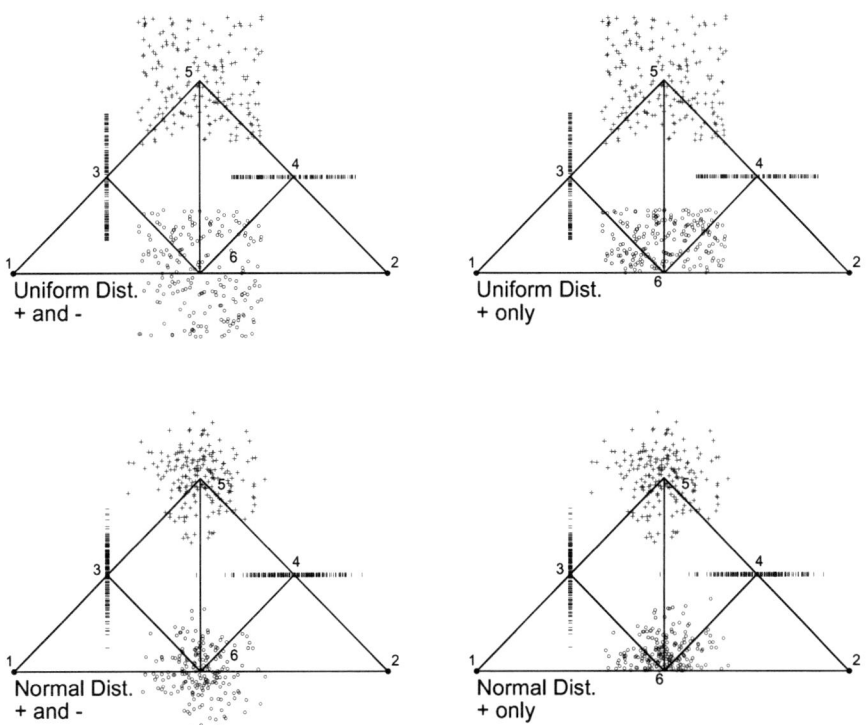

Figure 2.24. The four options for node mutation at restart.

Finally, only node coordinates which are neither marked as being neither supported nor held stationary, are allowed to mutate. In the example shown in Figure 2.24., the x and y coordinates node 1 and the y coordinate of node 2 are marked as supports, while the x coordinate of node 2 and 3, and the y coordinate of node 4 are marked as stationary. The x-y origin was taken at node 1.

Figure 2.25. shows the effect that increasing the level of choke has on the distribution of new random nodes around the original node. The choke level begins as a sixth of the span, and is choked down to one sixtieth in ten steps. The distribution of random nodes is thus determined by the scale of the structure independent of specific units of length.

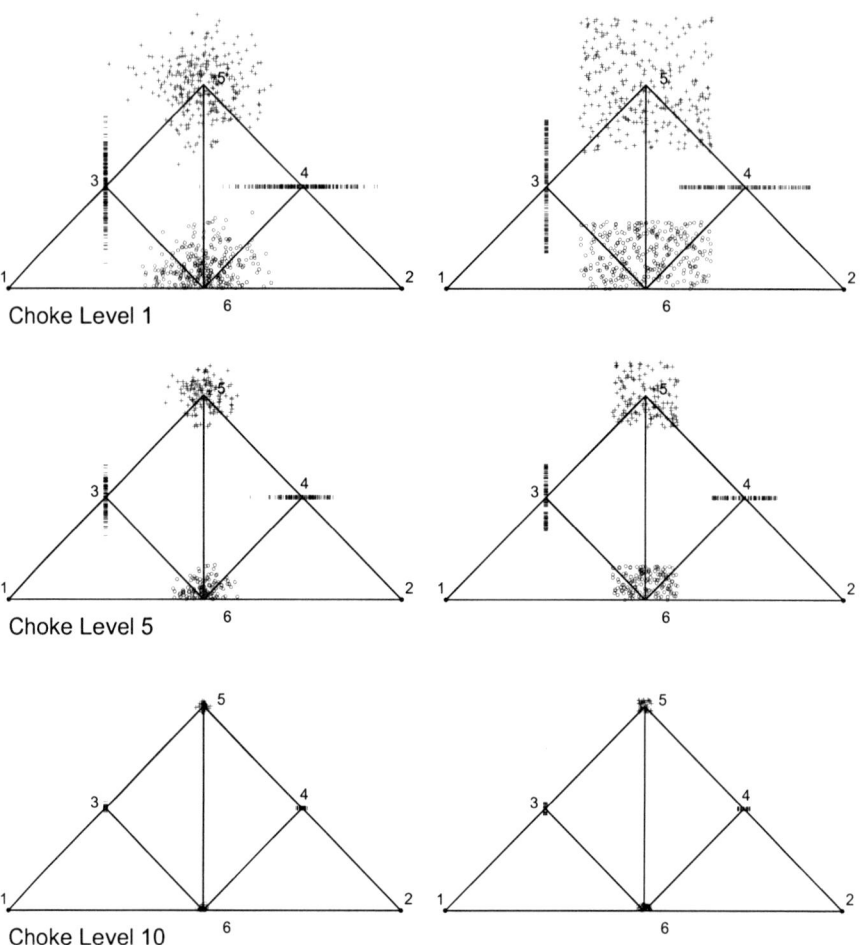

Figure 2.25. The effect of the progressive increase of choke level on normal sampling (left) and uniform sampling (right) as seen in three of the ten levels.

2.2.3.2 Run GA

The GA is the major portion of the cycle, and is where the geometry converges on the fitness function. It makes use of Eshelman's CHC-GA. Although the GA used is patterned after the CHC-GA, several modifications have been made to better accommodate the particular aspects of searching structural geometry and the use of real numbers. The modifications include:

- parent difference determination
- convergence recognition
- breeding with real numbers
- mutation throttle
- restart options
- cycle termination

Parent difference determination is the first step in the GA after a restart, or initial random start, as described above. Rather than a difference based on the Hamming distance as used with binary strings, a difference based on real geometric node variance of the two parents is used. As shown in Figure 2.6., the sum of the distance between corresponding nodes of two parents is used as a measurement of the difference. The higher the number, the more difference is exhibited by the pair. In order for breeding to take place the difference must be above a threshold. The threshold is initially set at an average value taken from the randomly paired population. As the population converges so that there is less difference in randomly paired parents, fewer children will be bred, which allows the generations to run faster. When fewer than 20% of the parents are producing children the threshold is decremented, which allows more parents to breed. A lower limit is needed on the threshold to prevent it from dropping to zero, and thus forcing all parents to breed.

Convergence recognition is based on the dropping number of children, and signals the convergence of the population. It is used to determine a restart. The two conditions which together signal convergence are: less than 10% of the population breeding, and the threshold already decremented to its lower limit. The plot of one cycle of breeding shown in Appendix B makes clear how the threshold insures parents with more difference are bred. This both enhances the thoroughness of the search as well as the speed of convergence.

Breeding with real numbers is also depicted in Figure 2.6. The technique developed is a blend of Half Uniform Crossover as used in the CHC-GA (Eshelman, 1991), and the mutational ellipsoids used in the ES-$(\mu+\lambda)$ Evolutionary Strategy (Bäck, Hoffmeister & Schwefel, 1992). The array of node coordinates (see Figure 2.4.) of each parent is used as the chromosome for breeding. First half of the node coordinates are randomly selected for crossover (as with CHC). The selected nodes are "crossed" by finding a new node within an area (similar to the ES mutational ellipsoids) surrounding the two joints. There is a choice of two routines to select the new, child node. The first uses a normal distribution for selection, and the second uses a uniform distribution. The routine with uniform distribution is more exploratory, while the normal distribution routine will usually converge more quickly. The two children are bred, using the randomly selected half set of the nodes.

Mutation throttle allows the amount of mutation produced at restart to be decremented with progressive cycles. In this way the early restarts result in the most disruption, and enhance exploration, while the later restarts are more refined and concentrate on a smaller area of the search space. The distribution of mutations is progressively choked as shown in Figure 2.25.

Restart options allow for additional flexibility in describing the range of the solution space or by focusing the mutation more narrowly by using lower σ values with normal distribution. These options are described in Section 2.2.3.1 above and shown in Figure 2.25.

Cycle termination is normally brought about by the population convergence as described above. There are however some instances when convergence may become stagnated in an area where geometric differences do not sufficiently affect the fitness. For this reason a limit is placed on the maximum number of generations allowed. Since topologies with more nodes can have more geometric difference than topologies with fewer nodes, the number of nodes is a factor of the generation limit.

2.2.3.3 Select

Selection of geometry, unlike topology, runs only in an automatic mode guided by the predefined fitness function. If this were not the case, the user would have to sift though many generations of forms that are truly unfit. This would serve little purpose, and would probably prevent the discovery of most good solutions. The selection employed, in keeping with CHC-GA is elitist. After breeding, the combined parent/child population is sorted by fitness, and the best are kept to become the parent population of the next generation. This strategy is also common in ES. The selection is further elitist in that the best individual is always maintained. In a restart, the best individual is used to generate the new population. In a re-initialization, although the population is randomly generated (as in an initialization) the best individual is maintained in one copy. This prevents loss of the best solution but at the same time allows adequate opportunity for other better solutions will be found.

2.2.4 Running the IGDT

Complete examples of runs and results are given in Section 3. This section describes some of the procedure used to setup and run those examples.

2.2.4.1 Initial Startup

Only two files need to be altered in order to describe a new problem: the input data file described in Section 2.2.1, and the header file, truss.h, described in Sections 2.2.2 and 2.2.3. An example input file, as well as the #define statements from the header file (truss.h) are given in Appendix A. If the header file is altered, the program will have to be recompiled. A make file is used to compile the five executables. The executables and input files are then distributed to the slave machines for access by PVM (Parallel Virtual Machine) the parallel message passing software (Geist et al. 1994). Various builds containing different output and debug messages are also controlled by #ifdef statements in the make file.

2.2.4.2 Output and User Interface

As described in Section 2.2.2.1, the user interacts with the program differently depending on the mode used. For an effective selection to be made it is necessary to be able to view the solutions together. When using the interactive mode with topology populations of a size that can be comfortably viewed on one screen (about 20 total, 10 parents and 10 children) a window is created in which the populations can be viewed. But with larger populations this may be a problem. If multiple monitors are not a possibility, simply plotting the entire population to paper output for viewing is a simple solution. The program can write AutoCAD files for each population produced, which can then be fully manipulated for viewing or plotting by the user. With the solutions in a CAD format, the designer has complete freedom to manipulate the geometries in a more convenient setting. The designer can alter solutions, plot them to paper, pin them to the wall, or carry them about. As described in Section 1.3.1.2, being able to manipulate graphic images in this way, can offer the designer effective creative stimulation.

When using the automatic program mode, several graphic output files are produced as well as more detailed text files. First plot files are made of each generation grouped by cycles. These are actually the same plots recorded in the interactive mode, only in the automatic mode they will be larger. In a long run these plots may contain 10's of 1000's of images which are not easily sorted through. Plots of this type are primarily useful in debugging and insuring the proper range of search is being made with out some unintended bias.

To make the results of the automatic mode more accessible there is a second level of output. This is a set of two files. The first is a file containing all new topologies generated. In this file reoccurring parents are removed. The second file is based on the aforementioned file, but has all duplicate topologies with similar geometries removed. In this file a lower limit of performance can also be set to restrict size still further. This file generally contains 100 to 200 solutions. Through use this was found to be a manageable number. The simple line images fit at about 100 on a 11x17 (DIN A3) sheet of paper. Visually scanning a couple of these sheets will give the designer a good insight into possible topology and geometry patters appropriate to the problem. Each solution is labeled with a few critical characteristics (weight, joint and member number) and has an ID number which referenced more detail statistics about the solution found in the accompanying text output. An example plot of this output is given in Appendix C.

2.2.4.3 Final Selection

As in any challenging design problem, the exploration of the solution space can continue over a wide range, and is usually only limited by constraints of time and available resources. The IGDT is intended to find many good solutions in a limited time. It is not, however, an exhaustive search, and can sometimes discover new areas of interest in the solution space, particularly when led in a different direction by the designer's selections during the interactive mode. It is a common technique in design methodology to allow a fallow period for idea incubation after an intense period of exploration. The IGDT accommodates this approach very well by allowing the designer to continue a previous session, by either starting with a new 'progenitor' solution, or inserting any stored solution into the current breeding population. In this way, the exploration of a problem may continue indefinitely.

3 Examples and Results

The following examples are chosen to demonstrate some of the potential of the IGDT, and by comparison with other published work show a verification of the results as well as the unique advantages that the IGDT offers the designer.

All of the examples are based on two-dimensional, flat truss configurations. That is, all joints are considered pinned with no bending moments present. This is not an inherent limitation of the IGDT, but rather a measure of initial expediency in programming this prototype version. The analysis component of the program was coded specifically for the IGDT to give greater control during development and testing. This FEA component can of course be expanded to include other elements with more degrees of freedom, or replaced entirely by and existing, more complete analysis package.

Nonetheless, there are already a great many analysis options present in this version of the program. The analysis has been incorporated as a set of modular subroutines that can easily be added to, or altered, to expand the functionality of the IGDT. Current analysis options include:

Member Design:
- solid rods - without buckling - simple P/A
- hollow pipe - with buckling - continuous sizes – following ASD US steel code analysis criteria (AISC, 1989)
- solid rods - with buckling - continuous sizes - following ASD US steel code analysis criteria (AISC, 1989)
- optional slenderness limits
- optional deflection limits

Loading Options:
- self weight
- nodal point loads (any number or orientation)
- moving point loads
- multiple load cases

Support Options:
- independent x or y axis support at any specified nodes
- independent x or y axis geometric fixing without support (i.e., nodes that are not relocated by the GA)

Geometric Limits:
- optional x or y limits on geometry space
- optional limits on number of members or joints generated

In addition to these options, there are other options which control the performance of the GA selection and breeding. These were discussed in Section 2.

From this list it is apparent that it is not practical to give examples of all permutations of options, but hopefully the examples that are presented will show the capabilities of the IGDT to an extent sufficient to demonstrate the concept.

Over the course of development and testing of the IGDT, several platforms and hardware configurations were used. Initial development at ILEK was possible on a single HP 9000 workstation, but with increased complexity and the addition of topology optimization, it was decided to employ parallel processing techniques. At that time a small cluster of 10 HP workstations was used. The following examples were finally run using a cluster of 30 Intel workstations – a mix of Pentium II's and III's ranging between 200 – 733 MHz. All run times cited below are for this last configuration.

3.1.1 Flat Deck Bridge

The source of this example is a small, two lane vehicular bridge. It is conceived as representing a flat deck roadway supported on either side by trusses. The scale of the bridge (span, load levels) has an impact on what forms are appropriate. Here the IGDT offers the designer the advantage of being able to tailor the exploration to a specific set of requirements.

3.1.2 Problem Description and Setup

3.1.2.1 Problem Parameters and Geometry

The panel segmentation was set to four panels to match a simple example by Noboru Kikuchi (et al. 1995). A panel spacing of 15 ft [4.53 m] results in a total length of 60 ft [18.3 m]. A 10.5 ft [3.20 m] lane width was used, which for two lanes results in a total width of 21 ft [6.4 m]. The three geometry constraints explored were:

1. truss above deck (pony or through truss)
2. truss below deck (deck truss)
3. truss both above and below deck (lenticular truss)

No limits were placed on the maximum height or depth of the trusses. All trials were made with members conforming to the design requirements of the AISC-ASD steel code (AISC, 1989) using continuous sizes of steel pipe. Figure 3.1. shows the three geometric configurations explored.

3.1.2.2 Loading Cases

The loading in all three geometric configurations was the same. The load cases considered were as follows:

- **DL** - actual self weight of steel trusswork (variable)
- **DL** - concrete deck at 225 psf (10.8 kN/m^2) (constant)
- **LL** - traffic at 640 plf [9.34 kN/m] HS20 lane load (constant)

Figure 3.1. shows the loading conditions. The self weight of the steel was calculated for the designed sections, and distributed to member end nodes. The design of the members was iterated, updating the self weight until 1% convergence was obtained for the sections. The deck dead load and lane load were applied as point loads on the deck nodes. A moving load was not included in this example to give better parity in comparisons with results from other researchers.

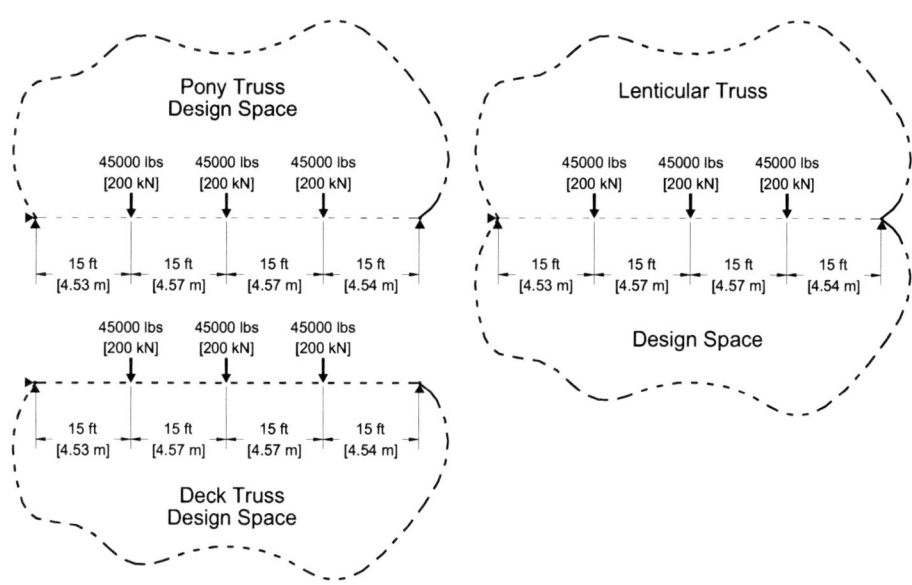

Figure 3.1. The Three Bridge Configurations Investigated.

3.1.3 Use of the IGDT

Although the basic geometry and single load case used in this problem are simple, the inclusion of the dead load does require additional iterations of the solution. Different startup methods were tried; random starts, progenitor individual and initial population. These are described in Section 2.2.2. The random start requires no initial geometry to be supplied by the user. In that sense it is an "unbiased" approach. Although it eventually reaches similar solutions that the other methods find, it generally takes more cycles to get there. The progenitor method, where one initial solution is supplied, is an expedient approach. The outcome is affected by the progenitor, but much exploration is still performed. In fact, as more cycles are run, the solutions often find various topologies regardless of the progenitor. Initial populations are similar to the progenitor approach, but require the user to supply more individual solutions. Many of these can be the same because the order in which they are listed in the input file determines which are bred with which. For example, with the four individuals, A, B, C and D, a population of 12 can be established with 6 unique pairs.

　　　AB　CD　BC　DA　BD　AC

The population size of 50 for a geometry generation was found to work well. This is a size recommended by Larry Eshelman (Eshelman, p. 303, 1995) and it was found to give good results without excessive computation. I occasionally use smaller geometry populations to get faster runs, but it does degrade the quality of the optimization. The size of the topology generation depends on the complexity of the structure. The complexity depends on the number of free joints being used. For this example and ones of a similar size, a population size of 20 was found to give good results. Higher numbers were tried, but did not seem to offer much improvement for the increase in run time. The selection criterion was primarily weight, but if deflection exceeded span/120, an additional sliding penalty was applied. As discussed in Section 2.2. there are several options in running the IGDT, either interactively or in automatic mode. The automatic mode makes selections based on pre-established criteria (in this case least weight). The interactive mode supplies the user with information like weight on each individual topology, but the user actually makes the selections for breeding at each generation. Since most of the examples, including this one, were to be compared with other optimized solutions, there was no advantage to run the IGDT interactively. Also, it is easier to use the automatic mode because problems can be set up and run unattended. Section 2.2 also describes the various termination options. A GA can run forever if you let it, but at some point there ceases to be any significant development. Termination methods include: set number of topology generations (with just one cycle), set number of topology cycles, automatic cycle termination after a set number of repeated final results. Of these three options the pre-set number of cycles was almost always used. Although as many as 30 to 40 topology cycles were sometimes run, more typically run lengths of 10 to 20 cycles were used. During the topology cycles, each time a new topology was generated, it was saved in a separate group which in the end was sorted by topology types (many are redundant) and the fittest from each type was copied into a new group which became the final output. This list could then be truncated using some limit of the fitness values. The final output then, was a list of trusses in an AutoCAD file that showed the user both variation of solution as well as good fitness to the criteria. In this way the IGDT successfully achieves the goal of disclosing a variety of good solutions to the user.

3.1.3.1 Results of Pony Truss

Figure 3.2. shows an example of a progenitor used to start the pony truss exploration. The initial cycles tended to have more variation, but less complex topologies with fewer members. In later cycles, the topologies increased in complexity, having more members. The greater number of members shorted the length between nodes. Since the nodes were taken to be braced, this gave the shorter compression members an advantage in buckling analysis.

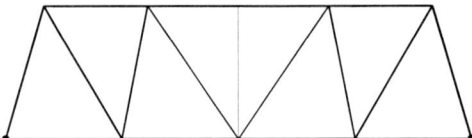

Topo ID: 0   weight = 2821 lb [1280 kg]

10 joints   17 members

Figure 3.2.   The progenitor pony truss used to start topology cycles.

Figure 3.3. shows selected topologies returned by the IGDT. As a means to understand the results, the topologies have been grouped based on some common feature. For example, 1, 2 and 3 all have a center upward peak /\ with a vertical member | at the center line. In the next set of three, 4, 5 and 6, the center peak is inverted \/ and there is no central member. Further the patterns have been arranged from left to right by

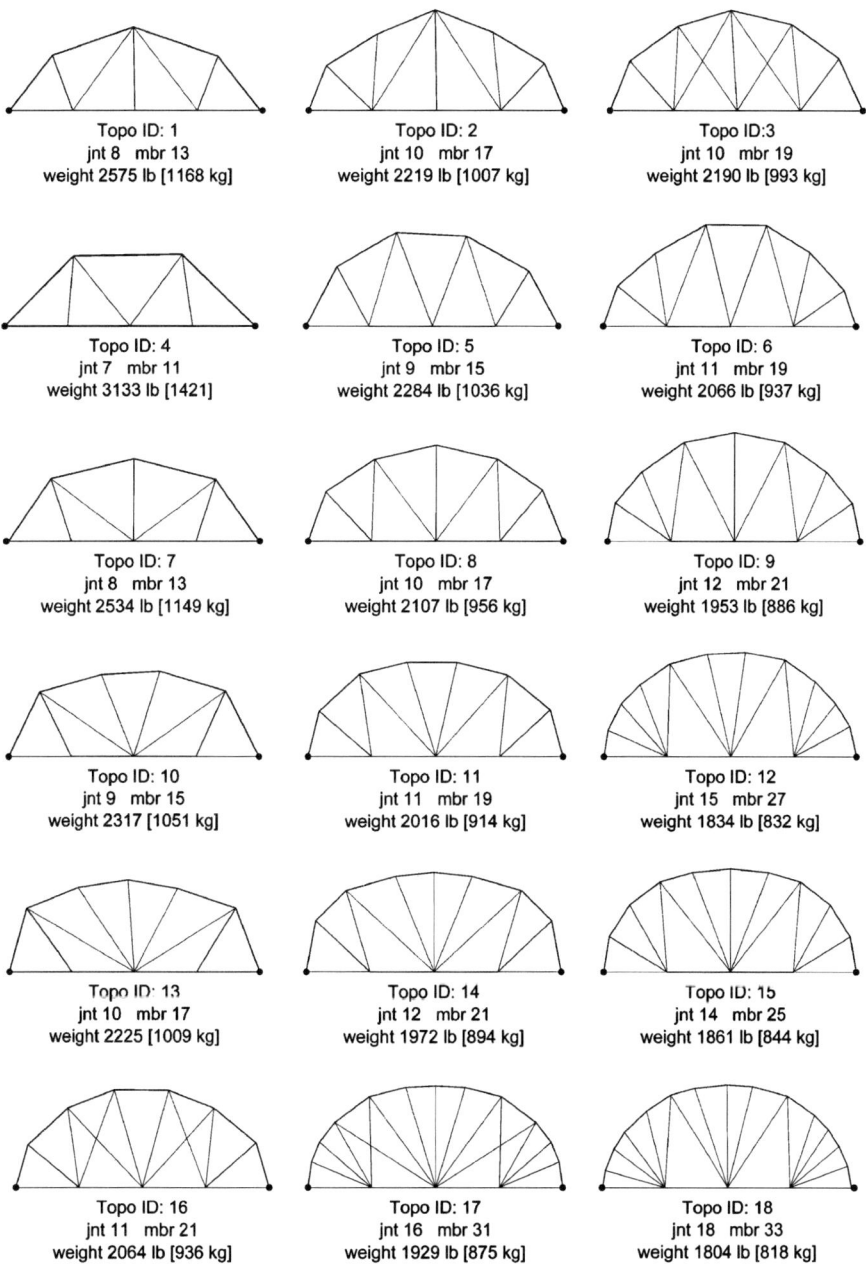

Figure 3.3. Selected results from the pony truss bridge. The results are reorganized here into topological families (rows) which gain in complexity from left to right.

increasing complexity. For example in sets 4 through 12 the condition at the center node remains constant for the set of three, while the condition at the side nodes increases from 1 (in topo 4) to 2 (in topo 5) to 3 (in topo 6). This is of course just my interpretation of the results and other users may see other groupings in exploring the output.

Finally, as pointed out by Boden (1994, p.76), Koza, et al. (1999, p.544), and others, one test for intelligence is the ability to discover solutions that are well known or patented. In looking through the list in Figure 3.3. names could be given to several. Topology 7 is a Waddell truss; topo 8 is a Parker; topo 5 is the Warren; topo 3 is a Bowstring Arch. That these topologies are successful enough to have names, indicates that the IGDT is finding good solutions.

A knowledgeable designer will often be able to detect critical parameters by observing the topology patterns. In the set shown in Figure 3.4. for example, Topo 18 has the least weight, even though Topo 20 has the same number of members. Apparently, the critical parameter is not just the number of members, but the relative lengths of the members in the compression arch. In Topo 18 the lengths are fairly even, shortening a bit near the reaction. In Topo 20 the arch lengths go from shortest in the center span, to longer, and then shorter again. The advantage achieved in Topo 18 is certainly due to the effect of buckling, and the fact that node points are considered braced. Of course it remains for the designer to judge whether or not the short lengths of the top chord are actually braced. This is a limitation of the current FEA routine. Since the analysis only considers flat plane trusses, out of plane bracing is not modeled, and simply assumed present at the nodes. This could be corrected by adding three dimensional elements and second order deflection iterations to the FEA routines. This is discussed further in the concluding chapter 5.

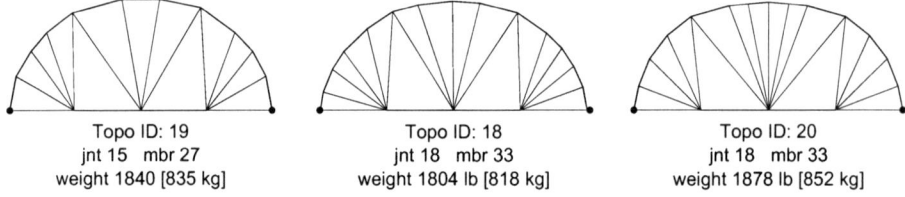

Figure 3.4. Three selected topologies from the pony bridge showing a pattern progression that indicates topology 18 as the most fit of that type. Both 19 and 20 have higher weights.

3.1.3.2 Results of Deck Truss

The procedure followed in exploring the deck truss bridge was similar to that used in the pony bridge described above in Section 3.1.2.1. The results have been sorted into pattern groups and are shown in Figure 3.5. Although the problem is basically the horizontal mirror situation to the pony truss, the results are very different due to the consideration of buckling in the compression members. The solutions found are generally less efficient, and a designer might consider allowing end supports and a compression arch below the deck level. But to maintain consistency in the set of pony, deck and lenticular trusses being investigated other support options were not explored.

In looking at the solutions a new strategy can be observed. Some of the fitter solutions are adding a second tensile arch as in topologies 12 through 15. Also, topologies 16 through 18 use a tensile member to subdivide otherwise longer compression members. Again, although this is in part an artifact of the assumed nodal bracing, the strategy is apparent and rational.

Figure 3.6. shows another topology progression of adding members that brackets a 20 member solution as being most effective in the set. Topology 18 also happens to be the fittest solution discovered in this run. But reviewing the other solutions one quickly sees that the advantage over others like topology 12 is not significant. In fact the true usefulness of the IGDT concept is seen not in the discovery of the 'best' topologies like 12 or 18, but in the finding of 'good' topologies like 8 through 10 or even 5 that are much simpler while only increasing in weight between 5% to 10%.

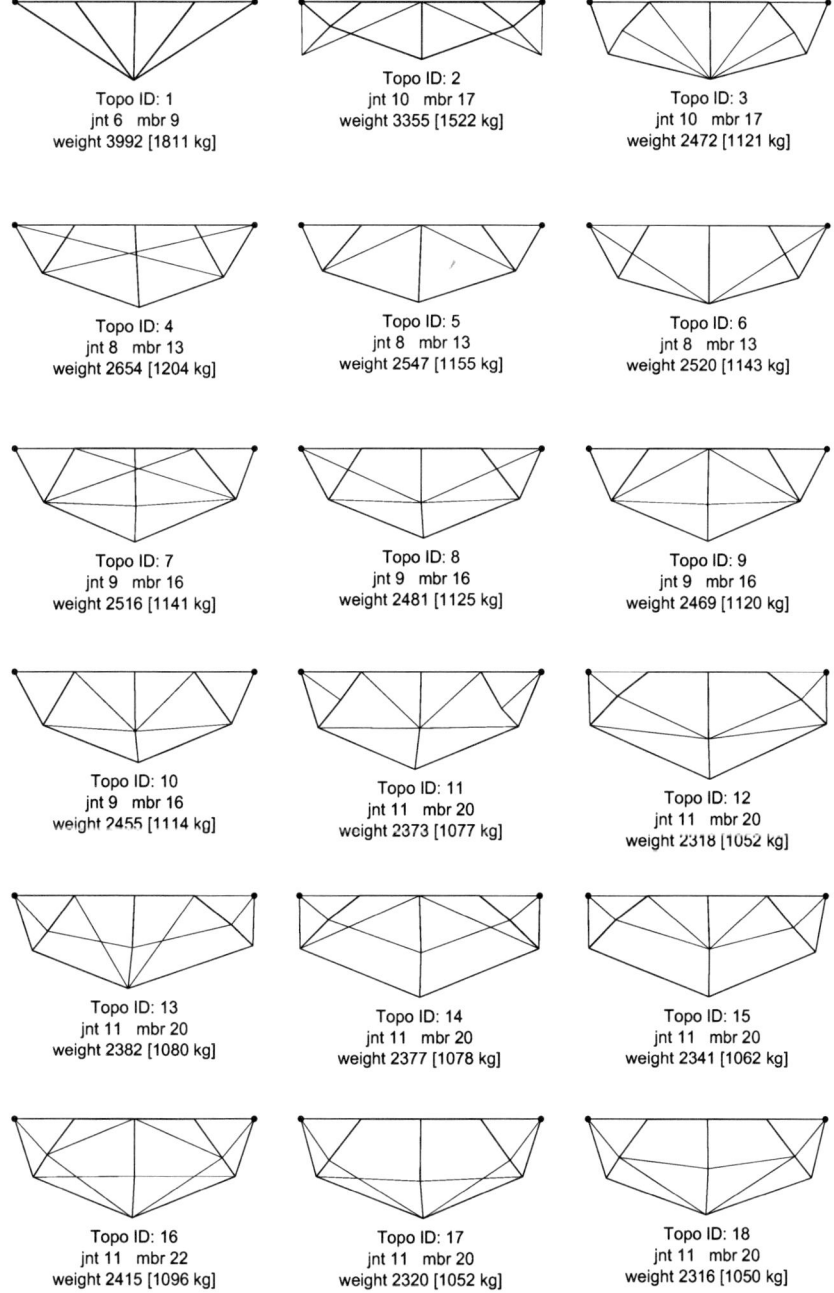

Figure 3.5. Selected results from the deck truss bridge. The results are reorganized here into topological families (rows) which gain in complexity from left to right.

As a designer looking for a practical solution, the discovery of 5 with only 8 joints and 13 members may be more useful than some more optimal solutions. Even if, through the use of some multi-objective optimization technique, you were able to find a solution like 5, with only that single solution, you would not have much of a sense of what else might be hidden in the solution space. At least with the IGDT you can see a selection like in Figure 3.5., and thus gain a better understanding of what other solutions are contained in that space.

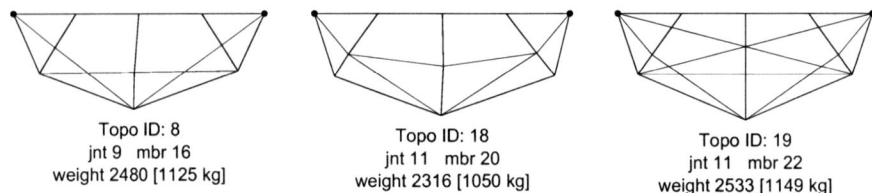

Topo ID: 8
jnt 9   mbr 16
weight 2480 [1125 kg]

Topo ID: 18
jnt 11   mbr 20
weight 2316 [1050 kg]

Topo ID: 19
jnt 11   mbr 22
weight 2533 [1149 kg]

Figure 3.6.   Three selected topologies from the deck bridge showing a pattern progression that indicates topology 18 as the most fit of that type. Both 8 and 19 have higher weights.

3.1.3.3   Results of Lenticular Truss

The last of the three variations of the truss bridge problem is the lenticular truss. This was analyzed in a similar manner to the other two described in Sections 3.1.2.1 and 3.1.2.2. The results are shown in Figure 3.7. below. Of the three variations, the lenticular configuration offered the greatest economy of weight. In this set several topologies were found between 1800 and 1700 lbs [820 – 770 kg.]. This compares to the lowest deck truss of 2316 lbs [1050 kg] and the lowest pony truss of 1823 lbs [827 kg]. Like the pony truss, the lenticular used tensile ties to subdivide the compression arch, and used far fewer compression struts below the deck. It is interesting too, to notice the relative height of the top compression arch as compared to the bottom tensile tie. Because of buckling, there is more advantage to be gained in reducing the force in the top arch members by moving them further from the deck. On the other hand, increasing the radius of the tensile tie below the deck requires longer compression struts which are more susceptible to buckling. Again, being allowed to see a selection of solutions is perhaps the greatest benefit offered by the IGDT. And again, the most practical solutions are not necessarily the solutions meeting the least weight criteria. For example topologies of 8 through 10 offer a good combination of economy of weight and simplicity of form. Topology 8 at 1809 lbs [821 kg] is only 5.5% heavier than the best performer 14 which weighs 1714 lbs [777 kg], and yet 8 uses only 26 members as compared to 30 members in topology 14. Or the even rhythm of three members above and one below at each deck node in topology 13 at 1746 lbs [792 kg] with 28 members, might be considered by the designer as more visually attractive, and a compromise of the aforementioned pair. Figure 3.8. shows topology 13 placed for comparison between 8 and 14.

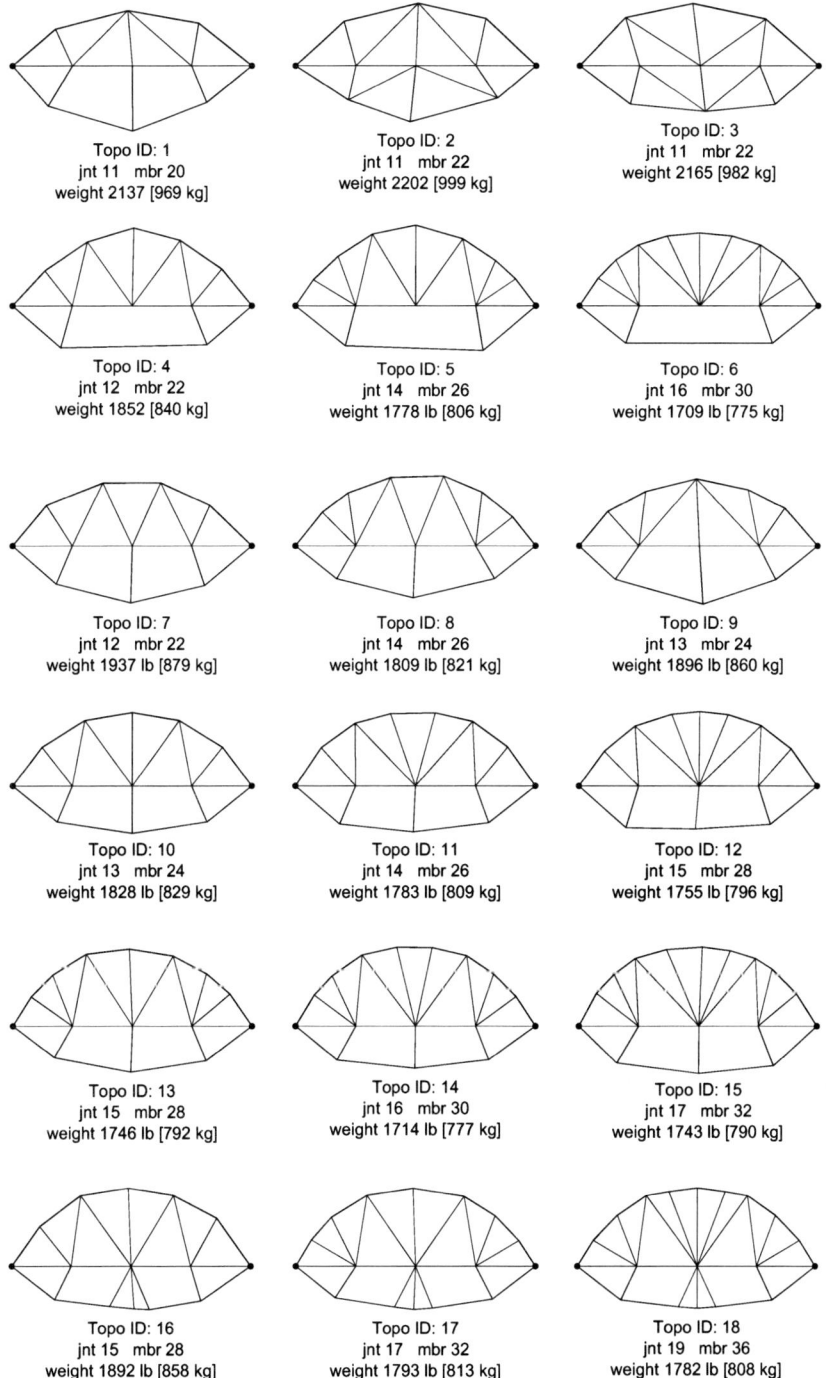

Figure 3.7. Selected results from the lenticular truss bridge. The results are reorganized here into topological families (rows) which gain in complexity from left to right.

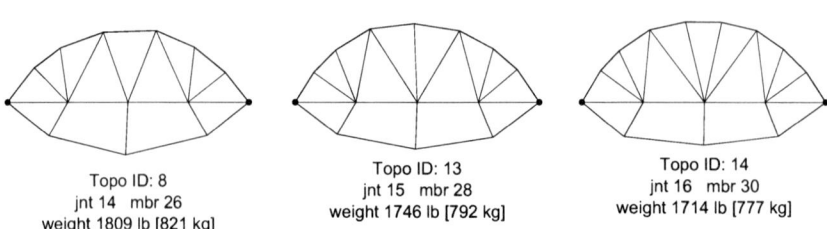

Topo ID: 8	Topo ID: 13	Topo ID: 14
jnt 14 mbr 26	jnt 15 mbr 28	jnt 16 mbr 30
weight 1809 lb [821 kg]	weight 1746 lb [792 kg]	weight 1714 lb [777 kg]

Figure 3.8. Three selected topologies from the lenticular bridge showing a range in the number of members and associated weights.

3.1.4 Comparison of Results

This section compares the results of the IGDT with other published results which have used either Evolutionary Algorithms or different optimization methods applied to similar small bridge problems as explored in this section. The first example shown in Figure 3.9. form Klarbring, Petersson & Rönnqvist (1995) is based on a ground structure. In a ground structure, a field of nodal points is defined in advance and the members are only allowed to connect to these nodes. Klarbring et al. uses a multi-objective variation of Linear Programming. The results thus obtained are very similar to what the IGDT offered in Figure 3.3. by the more fit solutions, e.g., topologies 12, 15 or 18. Because of the three node support, the Klarbring et al. solution shows a second and partial third compression arch. But as a helpful design tool, the Linear Programming approach used by Klarbring et al. seems to fall short of what the IGDT produce in Figure 3.3. Although the Klarbring et al. solution points to one of the same topology patterns as found by the IDGT, the LP method lacks the capability to reveal other near optimum ('pretty good') solutions. As a result the upper half of Figure 3.3., which contains most of the 'named' solutions, remains obscured to Klarbring et al.

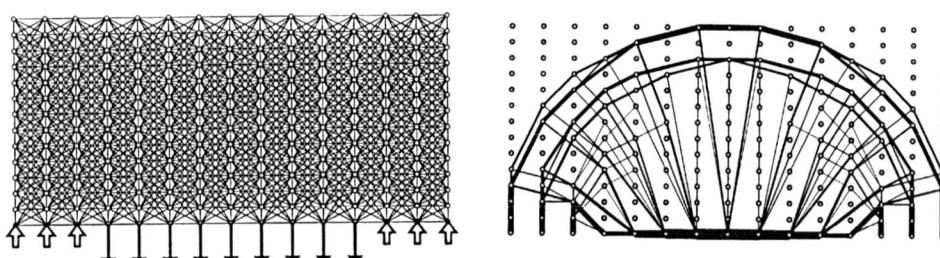

Figure 3.9. Left: Ground structure from Klarbring (1995)
Right: Ground structure nodes and optimized topology from Klarbring (1995)

Kocvara & Zowe (1995) provide a further comparison using Linear Programming both with and without a ground structure (Figures 3.10. and 3.11.). Figure 3.10. shows the solution derived using a ground structure with 264 nodes. Figure 3.11. shows a second solution from Kocvara & Zowe using the same design parameters, but this time no ground structure. The ground structure version can be seen as very similar to the example by Klarbring et al. in Figure 3.9. In Figure 3.11. the deck was held in place at set nodes and the arch was allowed vertical change of position. In the IGDT example both x and y axis position change was allowed, but either constraint is available to the designer. The arch in the Kocvara & Zowe example is a smoother parabolic form due to the increased number of load providing deck nodes.

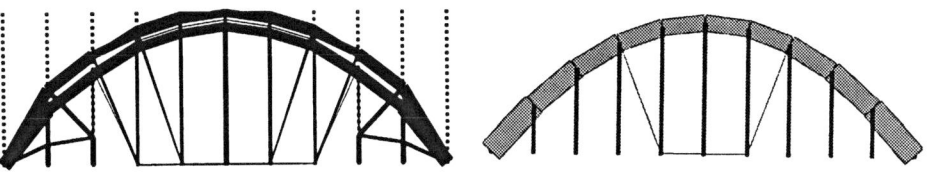

Figure 3.10. Ground structure and optimized topology from Kocvara & Zowe (1995)

Figure 3.11. Optimized topology without ground structure from Kocvara & Zowe (1995)

A third comparison is made with a pair of solutions published by Kikuchi, Cheng & Ma, (1995) based on the homogenization design method using a sensitivity analysis. In this example the topologies of the bridge trusses were optimized for given eigenfrequencies and weight. The results are a little hard to compare since I did not use the eigenfrequencies in my calculations. Also, the load distribution is different with the center load set at twice the level of the end loads. But Figure 3.12. is nonetheless interesting as an example of the output from the homogenization design method. This solution is a first order analysis. The gray scale indicates the level of stress in the members, with blacker being higher stress.

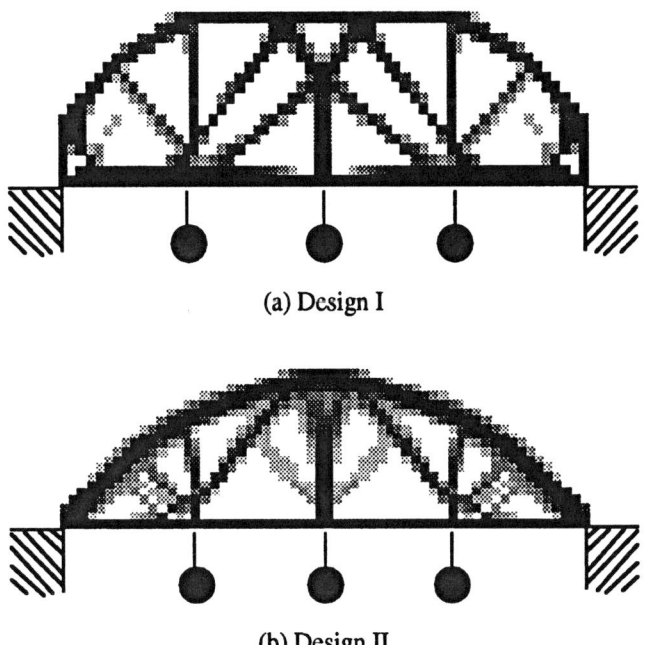

(a) Design I

(b) Design II

Figure 3.12. Two bridge topologies optimized for eigenfrequencies using homogeneous design methods and sensitivity analysis, from Kikuchi et al. (1995). Design I is optimized for 80, 170 and 200 Hz. Design II is optimized for 110, 130 and 170 Hz.

The last two examples were performed using stochastic methods. Figure 3.13. by Orta-Rial (2000) is an example of a lenticular bridge configuration. The method used was simulated annealing. This example, like the IGDT, is intended as a tool for the beginning phases of design. In that sense it searches for general forms. The truss topologies were optimized for weight using both permanent and variable loads at the deck nodes. The results are perhaps not so immediately applicable to a specific truss design, but do at least give some indication to the designer of load paths in the structure.

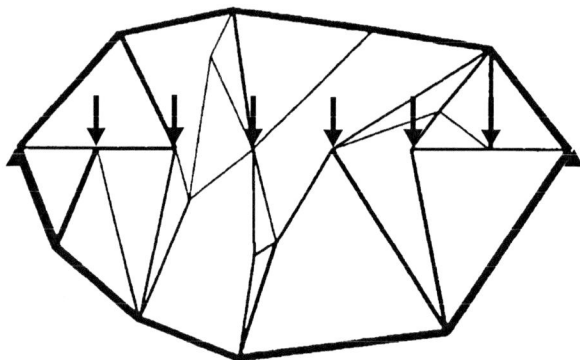

Figure 3.13. The topology of a lenticular seven panel truss, found using simulated annealing (Orta-Rial, 2000).

The last comparative example is by Deb & Gulati (1999) and uses a Genetic Algorithm to produce optimal geometry and topology for a bridge trusses based on a ground structure. Deb comments that most examples of truss topology optimization which use GA's find first a topology while holding all member cross sections constant, then once a topology is chosen the members are sized. Of course, in using a ground structure, the selection of topology and geometry are linked. That is, when a topology is chosen from the ground structure, the geometry is not altered from that ground structure. These two points certainly limit the quality of the results obtainable. Deb corrects the short coming of first point by optimizing each stable topology found to base the fitness on actual member sizes. But like all methods based on ground structures, he is limited by the second problem of linked topology and geometry.

The effect of linking topology with geometry is made apparent in Figure 3.15. On the left of Figure 3.15. is shown the topology/geometry found by Deb, based on the ground structure shown in Figure 3.14. Using the same loading and topology I employed the IGDT to search for a better geometry. The result is shown in comparison to Deb's solution, on the right of Figure 3.15. This comparison highlights the limitation of the use of ground structures. The IGDT avoids the problem Deb describes by optimizing member sizes for each topology, but in addition, the IGDT also goes beyond Deb's solution by avoiding the use of ground structures altogether. This demonstrates one of the unique features of the IGDT in the field of topology optimization of trusses, viz. no limiting use of ground structures.

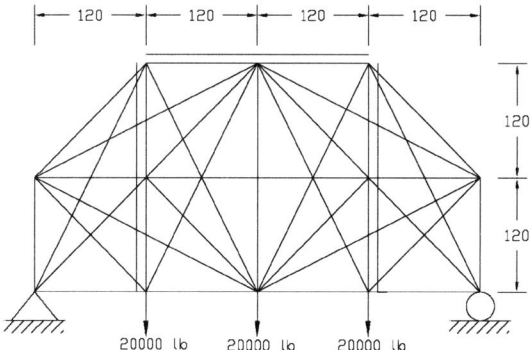

Figure 3.14. The ground structure used by Deb & Gulati (1999) for the truss shown in Figure 3.15.

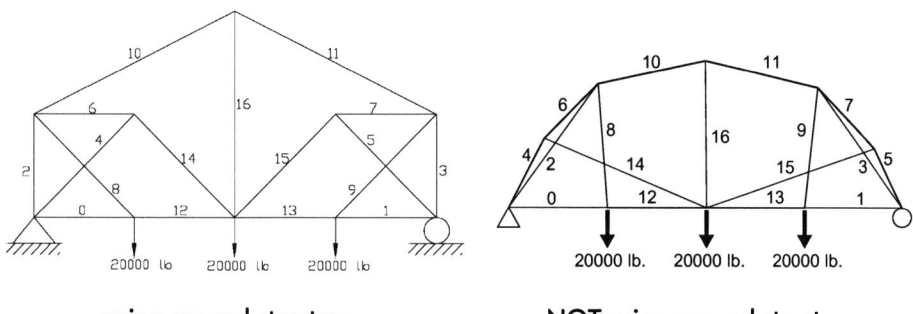

      using ground structure   NOT using ground structure

Figure 3.15. Two solutions based on the same topology. On the left is the solution found by Deb & Gulati (1999) using the ground structure from Figure 3.14. On the right is the IGDT solution found without using a ground structure.

Another aspect of ground structures is the effect the resolution has on the solution. Compare Deb's results using low resolution (12 nodes) with those of Klarbring in Figure 3.9., which use the much higher resolution of 210 nodes. Compare this with the results of the IGDT in Figure 3.4. As the resolution of the ground structure increases, the solutions become more similar to those of the IGDT. Of course the reason Deb chose such a low resolution has to do with the GA coding. The length of the chromosome is based on the total number of possible members in the ground structure. In Deb's ground structure this number is 39 as seen in Figure 3.14. The chromosome coding of the IGDT is based on the upper triangular portion of the incidence matrix. That number equals (joints^2-joints)/2 or in this case $(10^2-10)/2=45$. If Deb were to increase the resolution of the ground structure to something on the order of Klarbring's 210, the number of possible members in his ground structure would go way up to $(210^2-210)/2 = 21945$. Since the population size in a GA is linked to the chromosome length an increase from 45 to 21945 would require a significant increase in the population size as well. Both the longer chromosome and the larger population sizes would slow the convergence of the GA to an unacceptable level and the size of the problem quickly becomes unmanageable. However, since the IGDT chromosome length is determined by the actual incidence matrix size and not the ground structure size, the chromosome remains 45 even when using real numbers (actually doubles) for joint locations. In other words, the IGDT gets much higher resolution than Klarbring, at about the same cost computationally as Deb's unsatisfactorily low resolution GA.

3.1.5 Conclusions

The flat deck bridge example was chosen because it is both commonly used as an example in literature, and because it is a very familiar application for a truss structure which is well known to everyone. A broad range of solutions is presented in various relationships with the deck (above, below, and both above and below).

With regards to historical, build examples, the IGDT finds several designs which are so well known as to have names. Section 3.1.2.1 lists some of these trusses. Had the number of bays not been limited to 4, it is fairly certain that other more complex trusses would have been found as well. The discovery of these known truss cases by the IGDT is referred to as "historical creativity" by Margaret Boden (Boden, 1994, p.77). Boden argues that when a machine or program discovers an already known solution to a problem, it is exhibiting the same creativity as the original discoverer, only at a different time. In any case, it shows that the IGDT does find good solutions.

When compared with other published results, the IGDT again demonstrates its ability to find good results. However, more importantly, the results provided by the IGDT offer a choice of good solutions, rather than the single 'take it or leave it' solutions provided by other published methods. In each comparison with published results, the IGDT not just one, but several, as good or better solutions.

3.2 Arch Truss

This problem is taken from a thesis completed at the Institute for Lightweight Structures (IL) by Boris Peter (Peter, 2000). It has been chosen because it represents a solution found by traditional (for the most part trial-and-error) design procedure. It is also a case in which a capable engineering designer put more than the average amount of consideration into finding a solution that was both structurally efficient and geometrically elegant. In Peter's original solution, the geometry of the top cord of the arch was an initial given. This top chord served the dual function of cladding and structure, and was to be of glass. The truss elements beneath the arch provided the necessary stiffening which allow the shallow arch to carry half and full load patterns. Although the original thesis dealt in more detail with connections between glass and the steel rods, this example will concentrate solely on the design of the geometry for the supporting truss elements.

3.2.1 Problem Description and Setup

Mr. Peter's solution made use of solid steel rods in the design. The solid sections were chosen to achieve the most slender members, rendering a more filigree appearance. The choice of cross sectional type has, of course, an effect on the appropriate topology and geometry. Solid cross sections will require geometries which favor shorter length members as opposed to designs using hollow sections. Solid steel rods were used by the IGDT as well for consistency of comparison.

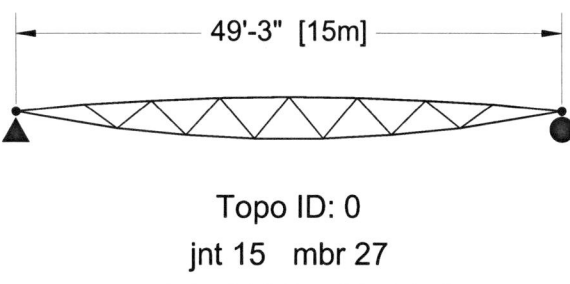

Topo ID: 0
jnt 15   mbr 27
weight 609 lbs [276 kg]

Figure 3.16.   Geometry used in Peter's truss "1c" (Peter, 2000).

Mr. Peter used, RSTAB by Dlubal, a commercial program incorporating a finite element analysis module and member design based on the DIN 18800, German steel code. Since the steel design routine currently used in the IGDT only designs sections based on the US steel code (AISC, 1989) the elements of Peter's structure were re-sized using the same AISC requirements. Both codes take member buckling into account. Member sizes ranged slightly higher in the AISC (allowable stress design) results, as compared to the DIN 18800 (strength design) results. Peter used 30 mm diameter sections for all members, and the AISC code gave a range of 20 mm to 40 mm. The AISC designed sizes were used in determining all of the weights used in comparing the different topologies.

3.2.1.1   Problem Parameters and Geometry

Figure 3.16. shows the geometry chosen by Mr. Peter. It is based on a familiar truss topology, commonly referred to as a Warren Truss. Complete geometric derivation can be found in Peter's thesis (Peter, 2000). The top cord was designed to have eight identical segments, each 6 ft. - 2 in. [187 cm] long. Therefore, the geometry of the top chord has been taken as a given for this example.

3.2.1.2   Loading Cases

Mr. Peter analyzed his truss using both symmetric and asymmetric loading patterns as well as out-of-plane loadings. Because the finite element routines in the IGDT currently contain only planar elements, the out-of-plane loadings were not considered. The loadings and load combinations that were considered are as follows:

Loading Cases:
- **DL** - self weight of steel truss plus glazing (constant)
  7.28 psf [0.348 kN/m^2] = 703 lbs [3.125 kN] per node
- **SL** - snow load on full span
  15.7 psf [0.75 kN/m^2] = 1515 lbs [6.74 kN] per node
- **SL½** - snow load on half span
  15.7 psf [0.75 kN/m^2] = 1515 lbs [6.74 kN] per node
- **WL** - negative pressure load (suction)
  -10.0 psf [-0.48 kN/m^2] = 967 lbs [-4.30 kN] per node
- **PL** - single point load
  2250 lbs [10 kN] per node

Load Combinations:
- **DL + SL** (Maximum downward load)
- **DL + WL** (Minimum downward load)
- **DL + SL½** (Eccentric half load)
- **DL + PL** (Extreme point load)

Figure 3.17. shows the six loading patterns applied to Peter's truss. The trusses were designed at a spacing of 15 ft - 9 in [4.80 m]. A detailed description of the applied loadings can be found in Peter's thesis (Peter, 2000). For purposes of comparison, the same load levels have been applied to this example. Although in the original design the top chord was a plate of glass and the bottom chord a cable, for purposes of exploring the possible topology of the supporting truss work, the entire system was modeled using steel rods. The geometry of the top chord (the glass arch) was taken as a given, and the topology of the supporting truss work was explored.

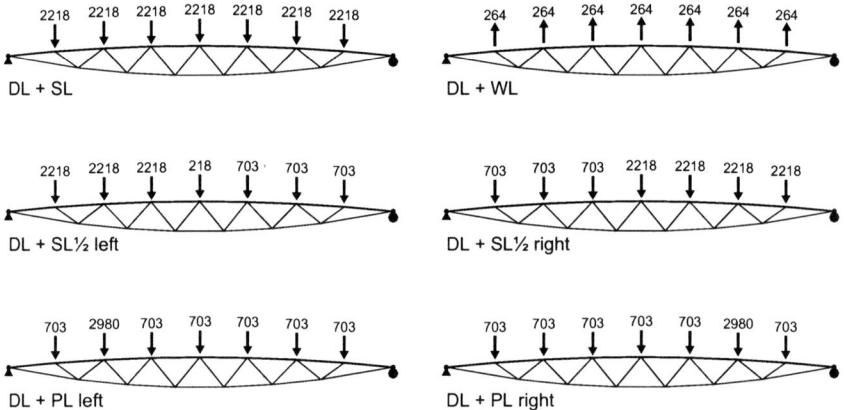

Figure 3.17. The six load combinations used for Peter's arch truss design. Nodal loads shown in lbs.

3.2.2 Use of the IGDT

As with the other examples, runs were made in the automatic mode to expose a group of plausible solutions. A selection of 18 solutions was made and grouped by topology type. In keeping with the original design intent, topologies having fewer members and shallower depth were sought. Figure 3.18. shows the selection of IGDT solutions. Finally, three topologies were chosen and forced to the same aspect ratio as the Peter truss for more direct comparison. Using the reversing load combinations shown in Figure 3.17., the top chord arch was always in compression and the bottom chord nearly always in tension. Topology 1 is an interesting exception being controlled entirely by compression. Web members tend to be either tension or compression, although compression dominated the member selection due to buckling with reverse loading. In general the simpler, statically determinate topologies without 'X' bracing were more weight efficient. The lightest solution found is topology 21 shown in Figure 3.19.

3.2.2.1 Results of Runs

The results of the IGDT runs showed several possible topology patterns. In the first row of Figure 3.18., forms with two or three nodes on the bottom chord with web members radiating from them are shown. The logical first pattern in this sequence, one lower node, was also discovered, but was not chosen for display due to the inefficient weight of 760 lbs. In the next row, topologies 5 and 6 use a bifurcating, tree-like support to the upper compression arch.

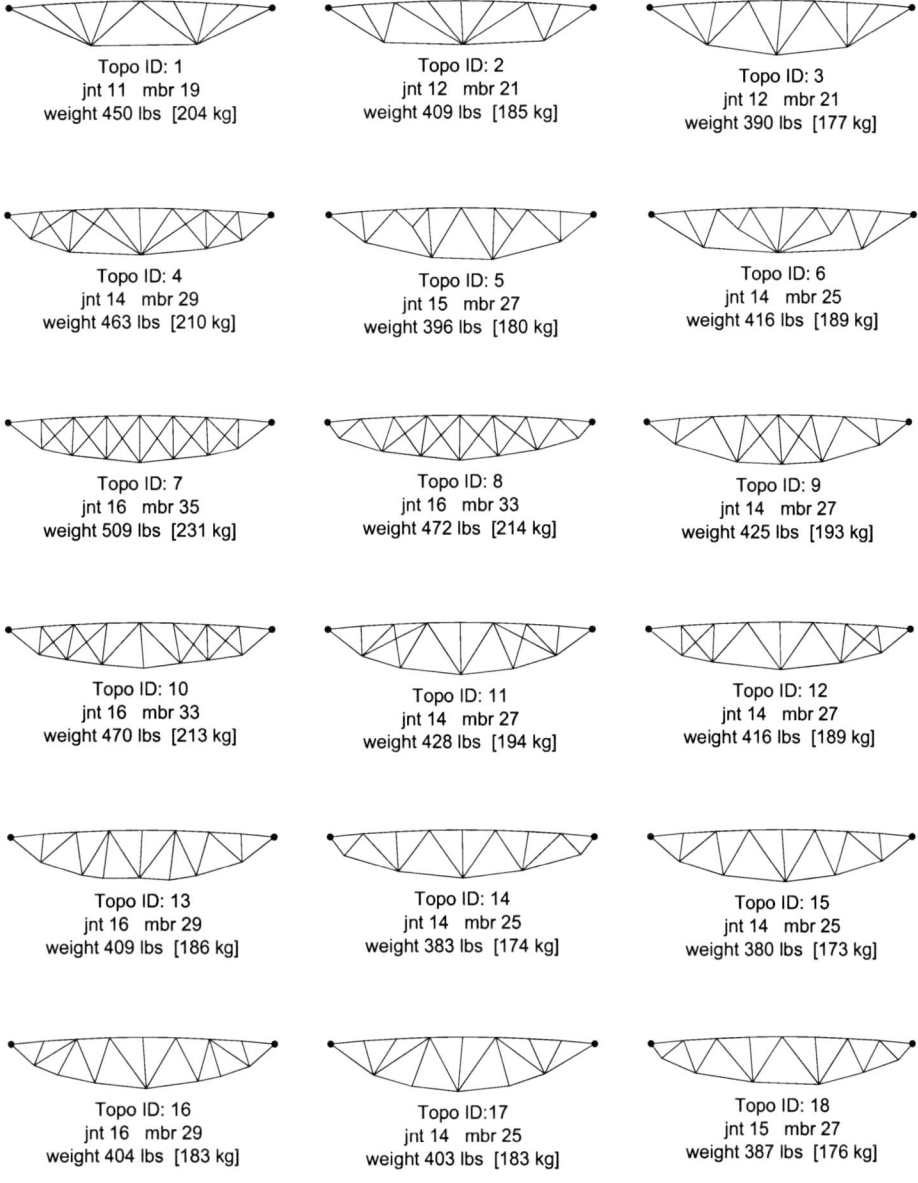

Figure 3.18. Selected results from the arch truss problem. The results are reorganized into topological families (rows) which gain in complexity from left to right.

3.2.3 Comparison of Results

All of the solutions found by the IGDT were deeper than the geometry chosen by Mr. Peter. The solutions shown in Figure 3.18. range in the number of joints from 11 to 16 and in number of members from 19 to 35. This range brackets the actual topology of the Peter truss with 15 joints and 27 members, shown in Figure 3.20. The weights of the IGDT solutions shown, which range from 470 lbs to 379 lbs, are all less than that of the Peter truss with 609 lbs. The large difference is primarily the result of the difference in aspect ratio, the Peter truss being considerably shallower. Figure 3.19. shows three topologies that were chosen for further comparison with the Peter truss. In the first row topologies 19 – 21 are shown with the optimal aspect ratio as selected by the IGDT. In the second row the same three topologies are shown using the same aspect ratio as the Peter truss. Even with the less optimal slenderness, all three topologies offer lower weights. The topology shown in 22 is actually the same topology used by Mr. Peter. In this case the IGDT was able to find a slightly altered geometry which offers a 5% weight reduction. The geometry in topo 22 is slightly thicker at the reactions, which presumably responds to the two peak point load cases shown in Figure 3.17. (DL+PL left and DL+PL right). The last geometry shown in Figure 3.19. (topo 24) is 6.5% lighter than the Peter truss. This was the lightest solution found by the IGDT.

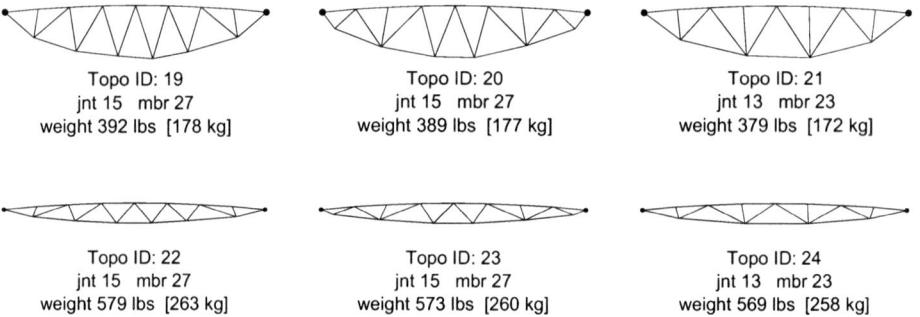

Figure 3.19. A selection of three topologies which best match the original selection criteria used by Mr. Peter. The first row shows the lightest geometries found by the IGDT and the second row shows the same three topologies reanalyzed with the same aspect ratio as used by the Peter truss.

Naturally the small advantage of lightness in these alternative solutions might be lost if a single cross-section were used for all members (as is commonly done). Nonetheless, it is apparent that the 1/2 span snow load and eccentric point load would be better accommodated by a less dramatically tapered geometry as Mr. Peter selected. In Peter's geometry, both top and bottom chords have a constant radius (the upper chord having about double the radius of the lower) which provides additional advantages in joint detailing.

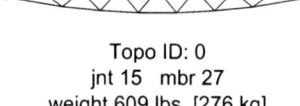

Topo ID: 0
jnt 15 mbr 27
weight 609 lbs [276 kg]

Figure 3.20. The original Peter truss geometry analyzed with the same member selection routine as used by the IGDT, in order to provide a valid comparison.

3.2.4 Conclusions

In the end, it must be admitted that the solution found by Mr. Peter with its regular geometry and smoothly tapered profile is more elegant than the slightly plumper IGDT solutions, and at very near the same weight. In this regard one can only compliment Mr. Peter on his refined sensitivity for structural form. At the same time, it must be admitted that the IGDT found not only the topology ultimately used by Mr. Peter, but several others that might have served as alternatives. Perhaps had Mr. Peter been presented with the IGDT solutions in Figures 3.18. and 3.19., he might have more quickly formulated his own design. Thus, this example illustrates that the suggestions made by the IGDT are in the same direction as those found by a capable engineering designer. In this way, it can be seen that the IGDT can function as a true aid in finding forms which lead to good engineering solutions.

3.3 Cantilever Truss

The cantilever truss with a single point load is a commonly chosen example in literature dealing with structural topology optimization (Bendsøe & Kikuchi, 1993; Galante, 1996; Hajela & Lee, 1995; Michell, 1904; Rajan, 1995; Rozvany, et al., 1992 Sankaranarayanan, et al., 1994). With stability is ignored, the problem was solved by Mitchell in 1904 using the method put forth by James Clerk Maxwell in the Royal Society of Edinburgh in 1870 (Maxwell, 1890). Mitchell's solution, as well as other solutions based on similar parameters, are shown in Section 3.3.3. in Figures 3.24. through 3.26.

3.3.1 Problem Description and Setup

Different researchers have used various aspect ratios and loadings in describing this problem. Often stability will be neglected in order to make a direct comparison with Mitchell's figure. In this example, two options were explored.

1. without buckling analysis (Mitchell case)
2. with buckling analysis using steel tubular sections

Both cases were otherwise run for the same load and geometric configuration. Figure 3.21. shows the problem setup and loading condition.

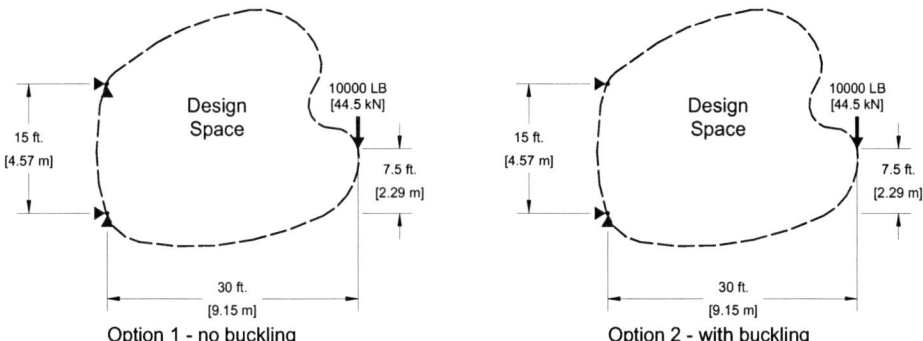

Figure 3.21. Description of Loading and Geometric Constraints for Options 1 and 2 of the Cantilever Truss Problem. Note the difference in end reactions.

3.3.1.1 Problem Parameters and Geometry

The geometries in each of the two options of this example were kept the same. Only the analysis parameter of buckling distinguishes the two trials. The dimensions are given in Figure 3.21. Steel was used in both cases and the AISC-ASD design specifications, including buckling, were followed for Option 2. Member area in Option 1 was determined based on a simple axial stress analysis (area = force / f_{yield}), i.e., buckling was not considered. The support conditions were pinned at both the upper and lower support. An overall aspect ratio of 1:2 was used. For option 1 without buckling, no limit was placed on size or slenderness of the members. Also, no penalty was made for amount of deflection. These parameters are similar to several published examples using other optimization methods (Duysinx, et al., 1995) (Kirsch, 1995).

3.3.1.2 Loading Cases

The loading for both options was also held constant. A point load of 10000 lbs [44.5 kN] was applied at mid height between the supports at the free end of the cantilever. Self-weight was not considered in either option.

3.3.2 Use of the IGDT

Both parts of this example were run in the automatic mode for a preset number of generations and cycles. Solutions were selected and arranged by topology and increasing complexity as in the other examples.

3.3.2.1 Cantilever without Buckling Analysis (Option 1)

Figure 3.22. shows a selection of 12 topologies found for the case without buckling. There are a few different patterns shown in the first two rows, but the well known Mitchell topologies fill the lower half of the figure. The process time for these runs was very short compared to the other examples. A 30 cycle run required approximately an hour for completion.

The actual difference in weight among the solutions shown in Figure 3.22. is not very much. This is an example of a solution plateau, where several solutions give approximately the same result. The advantage of the IGDT is that one is not only made aware of this fact, but also shown some of the other near optimal solutions as well. Using the IGDT in this way can help the designer obtain a feel for the sensitivity of the topology to variation.

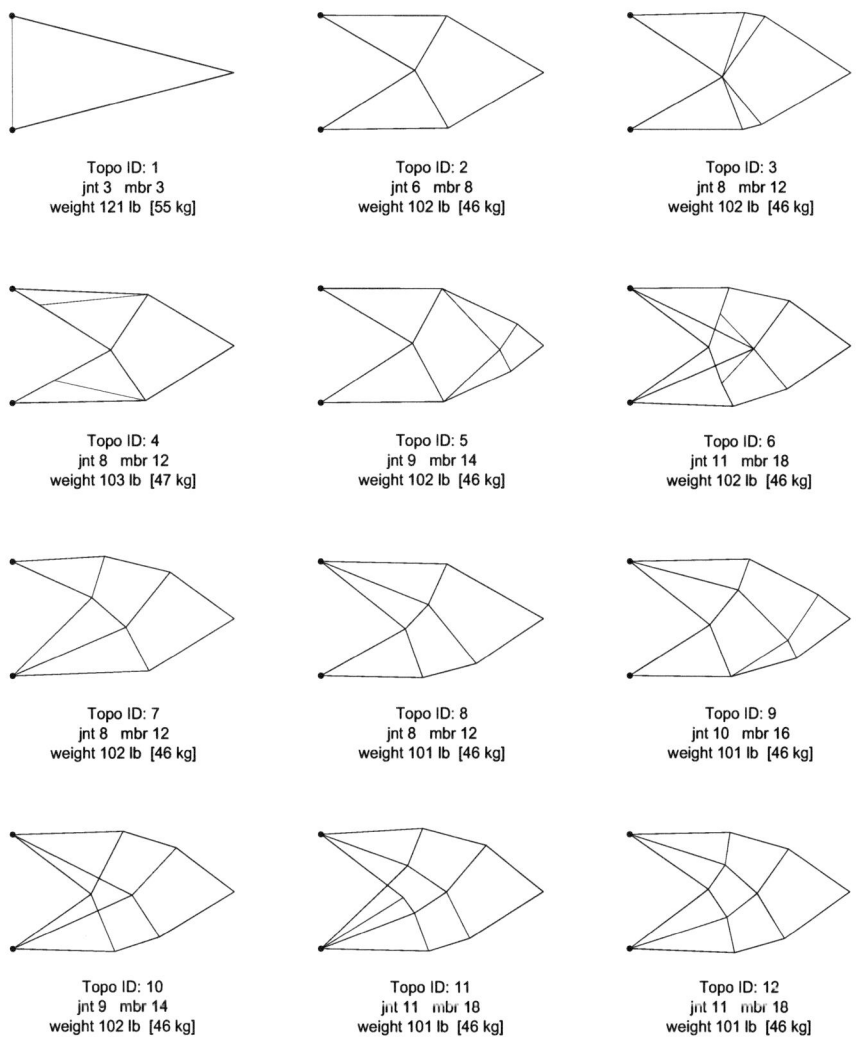

Figure 3.22. Solutions found by the IGDT for the non buckling case.

3.3.2.2 Cantilever with Buckling Analysis (Option 2)

The same cantilever problem was repeated with the members sized using the buckling criteria of the ACSA-ASD steel code (AISC, 1989). Members were chosen from continuous pipe sizes that approximate schedule 40 steel pipe. Figure 3.23. shows a selection of solutions found by the IGDT. As in each example the solutions are arranged with similar topologies three to a row, increasing in complexity from left to right as well as top to bottom. Here there was considerably more variation than in the non-buckling analysis. The familiar Warren topology is shown with increasing numbers of panels in the first two rows. Of this set, Topo 4 has the least weight. Atypically, the weight actually increases as more members are added from Topo 4 to 7. The topologies in the last row show some similarity with the Mitchell patterns, but in general the solutions were markedly different from those discovered in the non-buckling analysis. The symmetry of the non-buckling solutions is missing in these solutions.

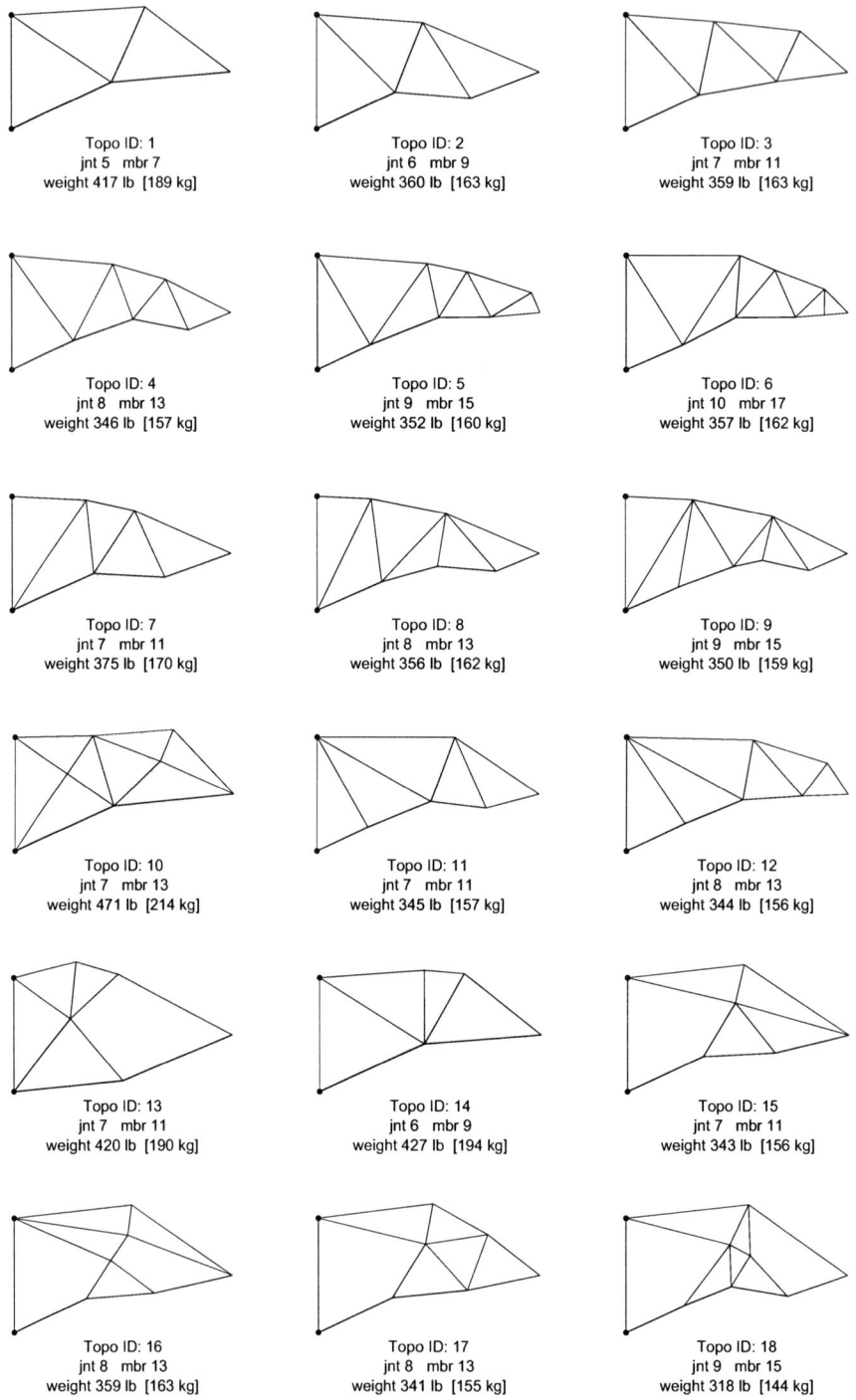

Figure 3.23. Solutions found by the IGDT for the buckling case.

3.3.3 Comparison of Results

This example was chosen mainly because it is used as a test case by may authors and therefore the results are readily comparable. The following two subsections show comparisons with different optimization techniques.

3.3.3.1 Cantilever Results without Buckling

The cantilever problem was run first without taking buckling into account, in order to be able to better compare the results with Mitchell's figure and other published solutions. Figure 3.24. shows the comparison of the IGDT solution to Mitchell's original cantilever example. The similarity to Mitchell's solution can be readily recognized. The IGDT topology is also comparable to the Klarbring solution shown in Figure 3.26. which makes use of a reduction strategy to find an optimal solution (Klarbring, et al., 1995).

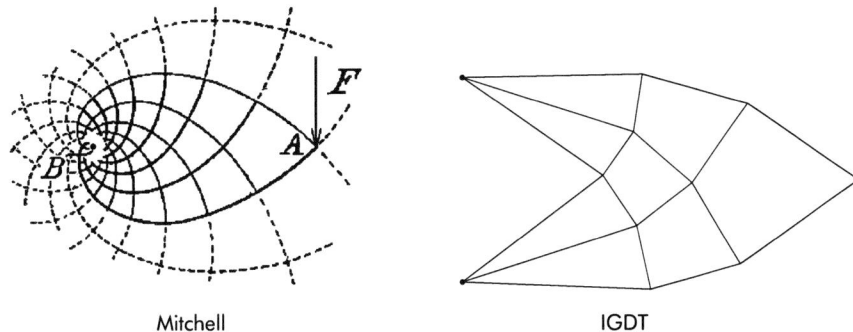

Mitchell   IGDT

Figure 3.24. A comparison of Mitchell's classic 1904 solution and the solution found by the IGDT.

Likewise, Figure 3.25. shows a comparison of a topology found using a homogeneous design method and sensitivity analysis (Duysinx, et al., 1995) with an IGDT produced solution. The topologies found by Kirsch (Kirsch, 1995) are very similar. Figure 3.26. shows the results found by Kirsch with comparable IGDT solutions. In all of these cases the IGDT succeeds in finding the same or very similar topologies. This can be taken as evidence that the algorithm is at least capable of discovering good solutions. But the real significance is not that the IGDT finds the 'best', but that it finds many 'good' solutions, and thus affords the designer a choice.

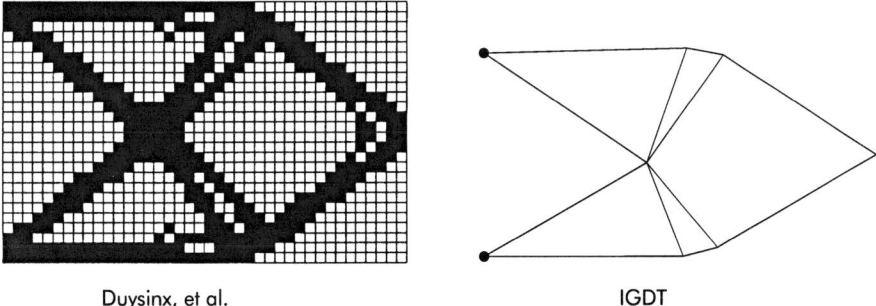

Duysinx, et al.   IGDT

Figure 3.25. On the left, optimization using homogeneous design methods and sensitivity analysis by Duysinx, et al. (1995). On the right, a topology found by the IGDT, Topo 3, Figure 3.22.

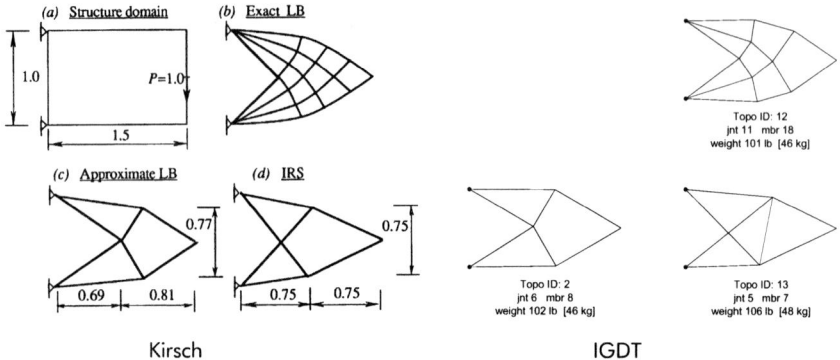

Figure 3.26. On the left, Cantilever topology optimization using a reduction strategy. (Kirsch, 1995) On the right, three similar solutions found by the IGDT.

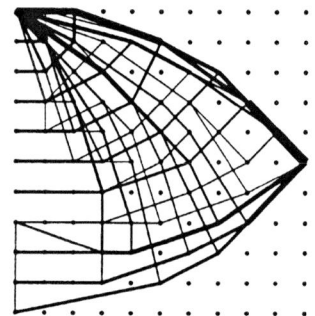

Figure 3.27. Use of a ground structure to find optimal topologies. (Klarbring, et al., 1995)

3.3.3.2 Cantilever Results with Buckling

Although I could not find published examples which included buckling, it is interesting to compare these solutions with those where buckling was not considered. In the runs made with the buckling analysis included, one can see the change in topology driven by the advantage of using shorter compression members and of placing a greater percent of the structure in tension. The current member sizing routine makes the rather simplified assumption that the members are always braced against buckling at nodes. This leads to many, short compression members as opposed to a few longer ones. Finally, the simple downward point load removes the chance of stress reversals and allows advantage to be taken by topologies that make use of slender tension ties. The top chord being in tension was usually less segmented than the bottom compression chord. For example in Topo 4 the top tensile chord is segmented into 3 members while the bottom compression chord is segmented into 4. The lighter solutions follow this pattern. Topo 8 is divided 3 top and 4 bottom, and Topo 9 is divided 3 top and 5 bottom. Topo 17 is the only solution under 359 lbs that does not have more top chord divisions than bottom chord. It is evenly divided 3 and 3. Another common attribute in the set is that the bottom chord is concave upward and the top chord is convex.

3.3.4 Conclusions

Again, this is an example which is commonly found in the literature. Starting with Mitchell's 1904 exact solution, many researchers have sought to verify their methods by reproducing the truss topology found by Mitchell. The IGDT is successful in finding this same pattern. The IGDT is also successful in finding patterns of other methods such as the homogeneous solution found by Duysinx, et al. The IGDT also has the flexibility to include buckling parameters and member cross section. The results found demonstrate the effects of including these additional variables.

3.4 High Speed Gantry

This is an actual design problem that was solved by an engineer using the finite element analysis and optimization capability of the commercial software package, ANSYS. It was chosen as a good example of how design problems are typically solved vs. results that can be achieved using the IGDT approach.

3.4.1 Problem Description and Setup

Of the examples presented in this dissertation, is the most complex in terms of usage and loading. Rather than a typical slowly moving gantry used to lift and move heavy objects, this gantry is intended to accelerate and decelerate a trolley at high speeds. It would be used to investigate the effects of different combinations of x and y axis acceleration and deceleration on the contents of the trolley. The trolley with its payload weighs 22 kips and the forces produced by its acceleration or deceleration are the primary loads on the structure.

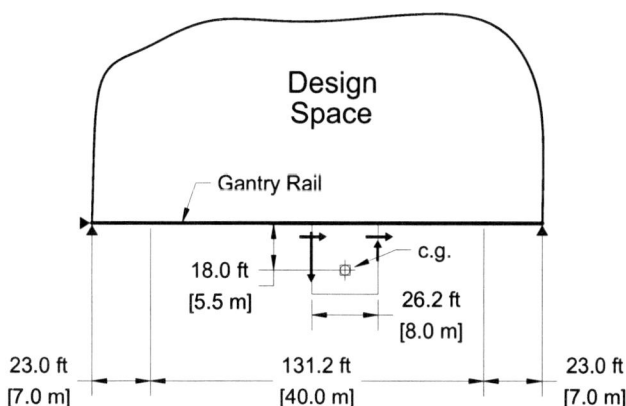

Figure 3.28. The gantry problem showing the location of the moving trolley. The trolley moves in this illustration in the horizontal direction across the 131.2 ft [40 m] span, while the gantry beam moves in the vertical direction of the illustration.

3.4.1.1 Problem Parameters and Geometry

The arrangement of the gantry and trolley is shown in Figure 3.28. In the actual problem, gravity acts normal to the plane of the drawing, but since the dominant loads come from the acceleration and deceleration of the trolley and gantry, the gravity load of the system is neglected. The weight of the loaded trolley is 22 kips [10000 kg] with a center of gravity shown in Figure 3.28. The trolley is prevented from coming any closer than 23 ft [7.0 m] to the ends of the gantry. In the orientation of the drawing, the gantry moves vertically, and the trolley moves horizontally. The gantry rail remains straight, and is supported by the gantry truss. Two basic arrangements for the truss were explored: six panel and seven panel.

3.4.1.2 Loading Cases

The problem has a complex loading caused by the acceleration and emergency braking of a trolley which moves on a track below a gantry. The load cases considered are taken from combinations of maximum accelerations and decelerations of the gantry and trolley. Because the trolley has a set width, the actual nodal loading depends on the panel spacing. The bottom chord (the track) was assumed to be a continuous member when determining the placement of the trolley to produce the worst case nodal loadings. The load cases were attained by using the highest pair of nodal reactions caused by the simultaneous braking of the trolley and gantry, while locating the trolley at each possible position and direction with respect to the panel nodes. Two panel configurations were explored – 6 panel and 7 panel. The load cases for each of these are shown in Figures 3.29. and 3.30.

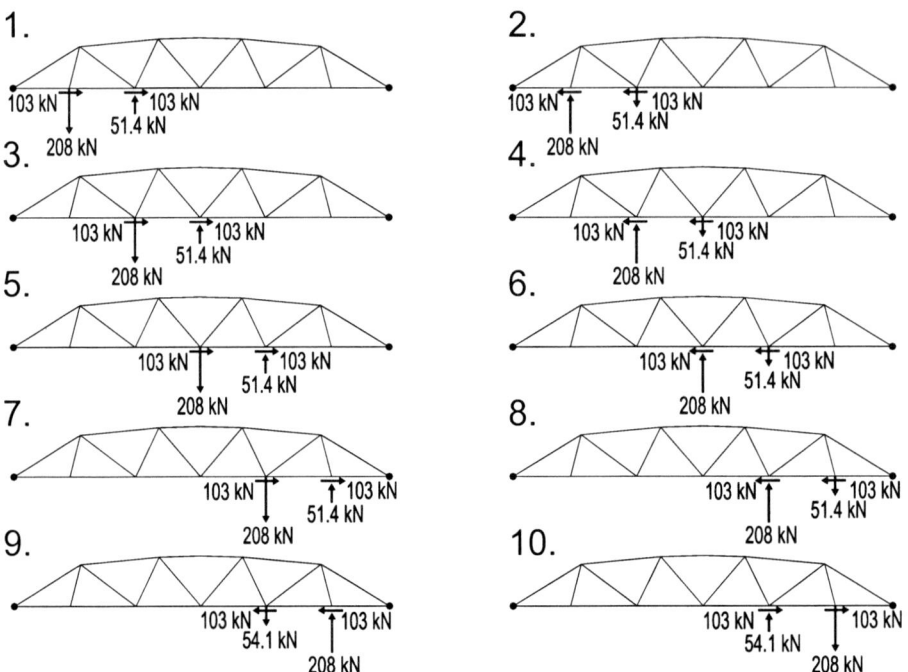

Figure 3.29. The 10 Load cases used in the six panel configuration of the gantry problem.

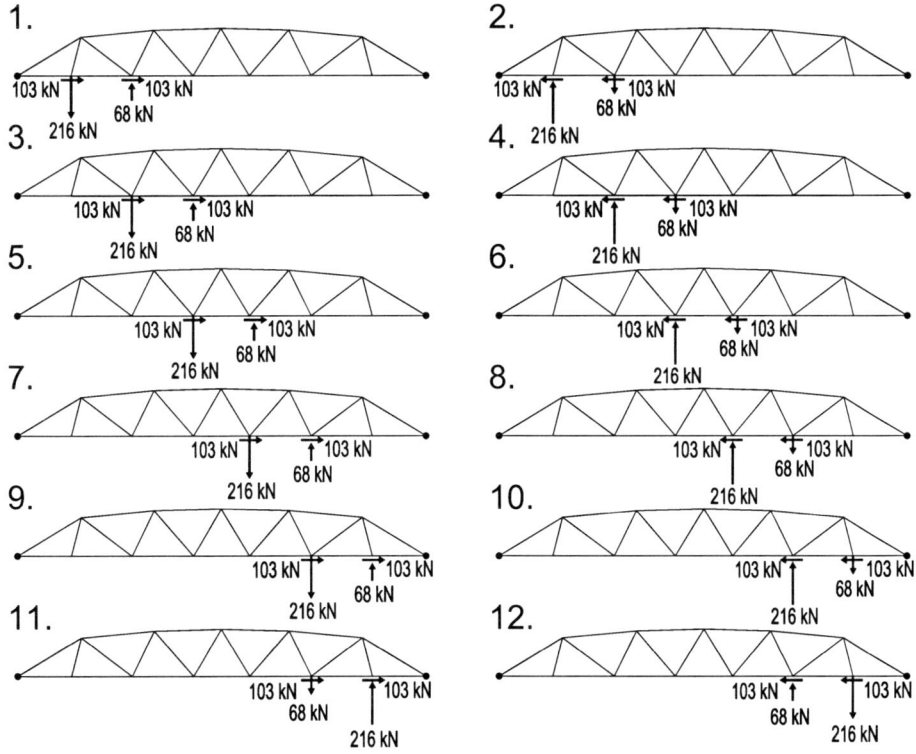

Figure 3.30. The 12 Load cases used in the 7 panel configuration of the gantry problem.

3.4.2 Use of the IGDT

Because of the complex loading and greater number of nodes, this problem took longer to run that any of the others presented in this dissertation. Using a cluster of 30 Pentium II & III class machines the run time averaged between 3 to 4 hours per topology cycle while using 50 individuals in the geometry generations and 20 individuals in the topology generations. Therefore, a complete run of 10 cycles took about a day and a half. A few different runs were made of each panel configuration using different startups (different progenitors or different start generations). After inspecting the results, it was seen that lower weight solutions were being found with more members. Because of the reversing load cases, almost all members are placed in compression under some loading, and therefore the shorter members have an advantage in buckling. In response to this, an additional run of 10 cycles was made setting the maximum joint count up to 32 and 36 for 6 and 7 panel trusses respectively, while the topology population was increased to 50. The larger topology population with more joints, caused the run time to increase to about 1 week. The solutions chosen from these longer runs are generally the lower half of the Figures 3.31. and 3.32.

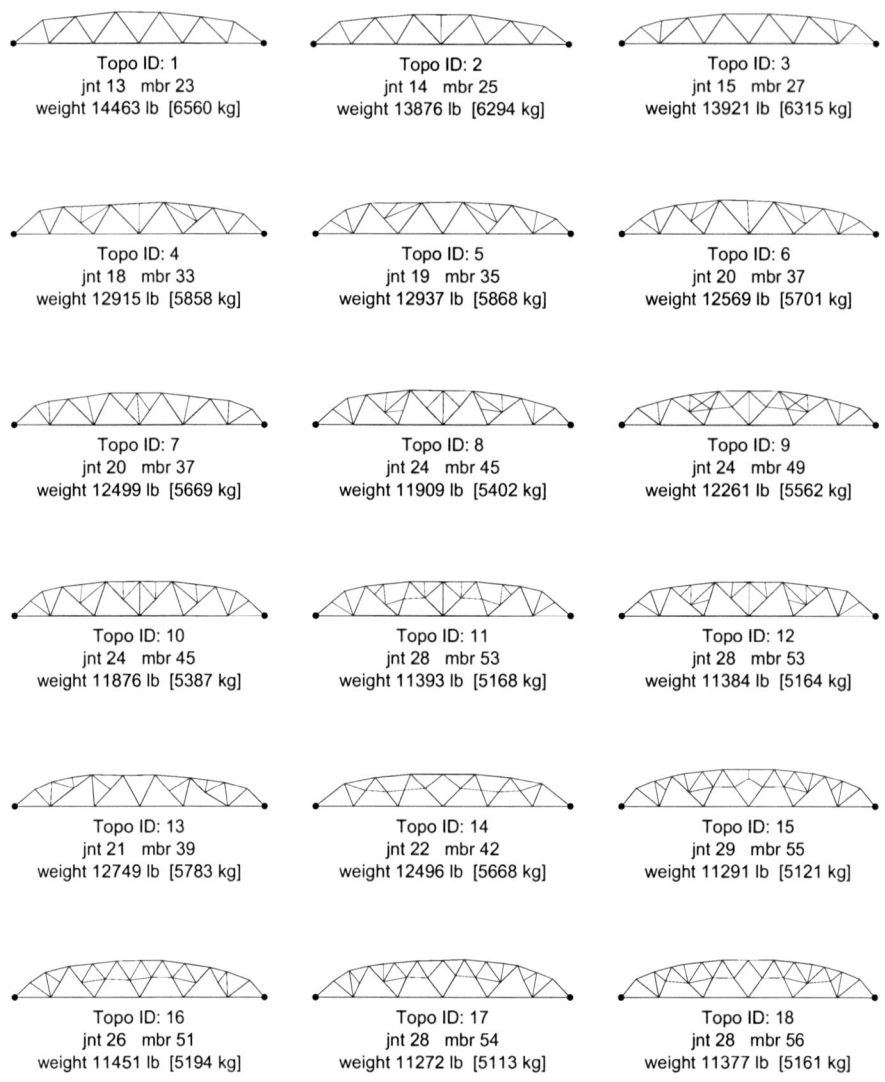

Figure 3.31. Selected solutions to the 6 panel configuration of the gantry problem.

3.4.2.1 Results of 6 Panel Truss

Selected solutions to the 6 panel truss are shown in Figure 3.31. Using 6 panels results in a panel width which is a bit wider than the trolley width (29.5 ft panel and 26.2 ft trolley). The results from the runs are arranged in topologically similar groups of 3 to a row with increasing complexity from left to right, as well as increasing complexity from top to bottom. Due to the reversing load cases, almost all member sizes are controlled by compression buckling. Therefore, there was an advantage in reducing member length of the inner members as well as the upper arch. The lower track member was preset, and therefore remains unchanged. This is certainly a drawback in the way the IGDT is coded. Because loading can only occur at nodes, the loaded nodes (and connecting members) have to be geometrically fixed. Therefore, the panel spacing on the loaded nodes is set. However, in most practical applications this may not be too severe a limitation. Solutions near the base of Figure 3.31. are about 20% lighter, but contain over double the number of members. So beyond a point, discovering new topologies with increasing numbers of members is not practical.

The simplest topology discovered (Topo 1) is the well known Warren truss. This had the least number of members of those shown in Figure 3.31. In this topology, the top chord is divided into 5 segments (regarding left and right members as 'sides'). In the ensuing variations, the top chord is increasingly subdivided – 6 segments in Topo ID 2, 7 segments in Topo 3, etc. In the last row the top chord has between 10 and 12 segments. What emerges is a subdivided Warren truss with the top half having twice the number of divisions as the bottom half. In all of the solutions, the aspect ration remains fairly constant at about 1:8.

3.4.2.2 Results of 7 Panel Truss

Selected solutions to the 7 panel truss are shown in Figure 3.32. Using 7 panels results in a panel width which is a bit narrower than the trolley width (25.3 ft panel and 26.2 ft trolley). The results are similar to the 6 panel solutions discussed above (3.4.2.1). Again the simplest solution in terms of joints and members is the Warren truss (Topo 1). As in Figure 3.31., the solutions in Figure 3.32. are arranged in by topology with complexity increasing from left to right and from top to bottom. The range in weight from heaviest to lightest is about 20%. In general the solutions of the 7 panel trusses where 2%-4% lighter than the 6 panel solutions. As with the 6 panel trusses, there was a noticeable strategy to subdivide the upper chord by adding a second row of smaller triangles.

As one might expect, as the complexity increases, the number satisficing, or sub-optimal, solutions also increase, and the sensitivity of the structure to minor topological variations decreases. Although most of the selected topologies exhibit bilateral symmetry, there were just as many, with similar or better fitnesses, that had at best imperfect symmetry or in some cases no visible symmetry. Figure 3.33. shows 3 of this category. With the larger number of intersticcs, the exact ordering of the topology seems less critical. The structures begin to take on a more organic appearance.

3.4.3 Comparison of Results

The IGDT results in Figures 3.31. and 3.32. can be compared with the design chosen based on the ANSYS analysis and shown in Figure 3.34. In order to make a useful comparison, the weight given for Figure 3.34. is based on the same pipe sizing algorithm used by the IGDT. Because ANSYS can only perform shape optimization on one load case at a time, separate solutions are obtained for each individual load case. Figure 3.35. shows two of these. It is then up to the designer to somehow combine the results or choose the worst case. One can see here the influence ANSYS had on the outcome of the original design.

On the other hand, the solutions offered by the IGDT represent the combined effects of all load cases. The IGDT solutions also consider buckling lengths of compression members. All of the IGDT solutions pointed to a flatter, less arched, form, with an aspect ratio of about 1:8 rather than 1:3.5 as used by the designer.

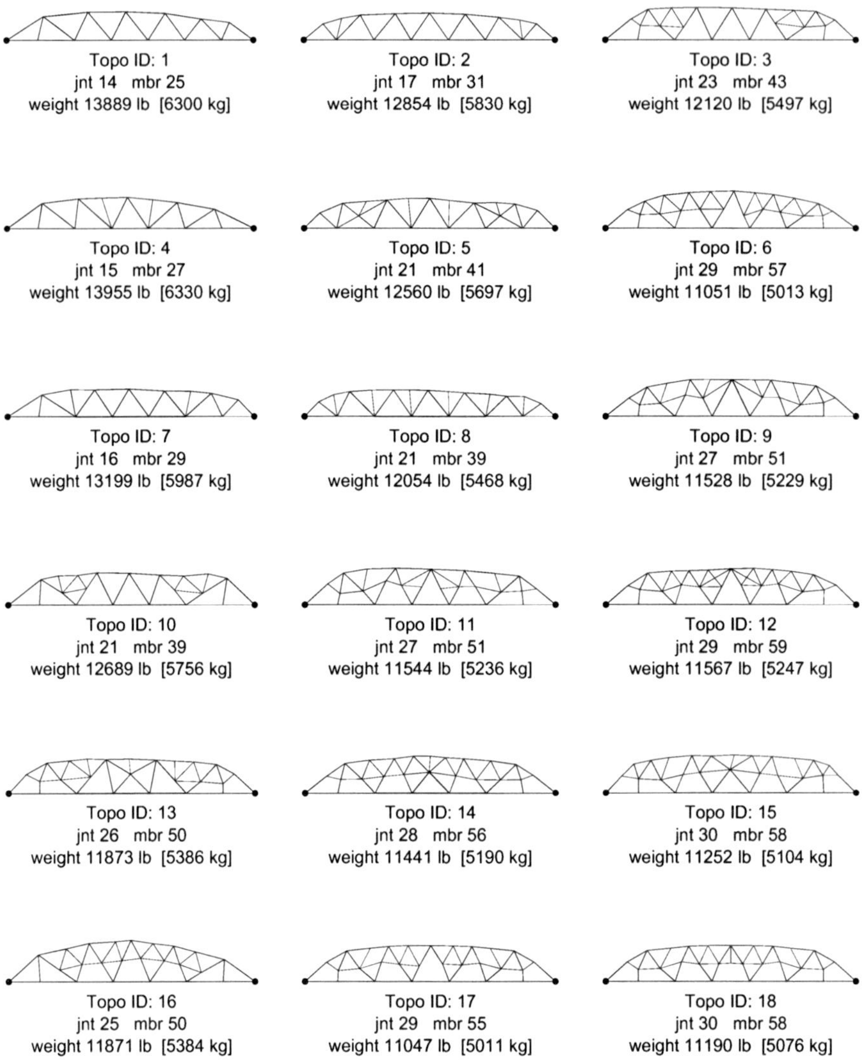

Topo ID: 1
jnt 14 mbr 25
weight 13889 lb [6300 kg]

Topo ID: 2
jnt 17 mbr 31
weight 12854 lb [5830 kg]

Topo ID: 3
jnt 23 mbr 43
weight 12120 lb [5497 kg]

Topo ID: 4
jnt 15 mbr 27
weight 13955 lb [6330 kg]

Topo ID: 5
jnt 21 mbr 41
weight 12560 lb [5697 kg]

Topo ID: 6
jnt 29 mbr 57
weight 11051 lb [5013 kg]

Topo ID: 7
jnt 16 mbr 29
weight 13199 lb [5987 kg]

Topo ID: 8
jnt 21 mbr 39
weight 12054 lb [5468 kg]

Topo ID: 9
jnt 27 mbr 51
weight 11528 lb [5229 kg]

Topo ID: 10
jnt 21 mbr 39
weight 12689 lb [5756 kg]

Topo ID: 11
jnt 27 mbr 51
weight 11544 lb [5236 kg]

Topo ID: 12
jnt 29 mbr 59
weight 11567 lb [5247 kg]

Topo ID: 13
jnt 26 mbr 50
weight 11873 lb [5386 kg]

Topo ID: 14
jnt 28 mbr 56
weight 11441 lb [5190 kg]

Topo ID: 15
jnt 30 mbr 58
weight 11252 lb [5104 kg]

Topo ID: 16
jnt 25 mbr 50
weight 11871 lb [5384 kg]

Topo ID: 17
jnt 29 mbr 55
weight 11047 lb [5011 kg]

Topo ID: 18
jnt 30 mbr 58
weight 11190 lb [5076 kg]

Figure 3.32. Selected solutions to the 7 panel configuration of the gantry problem.

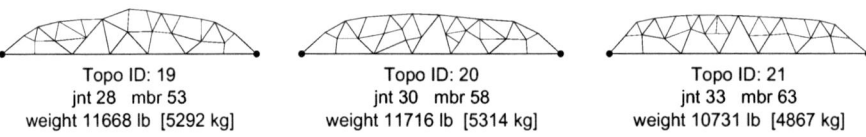

Topo ID: 19
jnt 28 mbr 53
weight 11668 lb [5292 kg]

Topo ID: 20
jnt 30 mbr 58
weight 11716 lb [5314 kg]

Topo ID: 21
jnt 33 mbr 63
weight 10731 lb [4867 kg]

Figure 3.33. Three examples of asymmetric solutions from the 7 panel run. Topo 21 was actually the most fit from the runs made, however, topo's 19 and 20 both have relatively good fitness values (low weight).

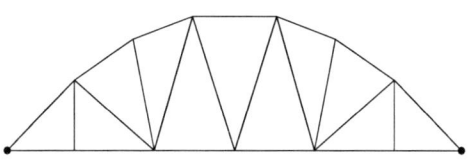

Topo ID: 0
jnt 13   mbr 23
weight 20472 lb [9286 kg]

Figure 3.34.   The final design chosen aided by a commercial optimization analysis (ANSYS)

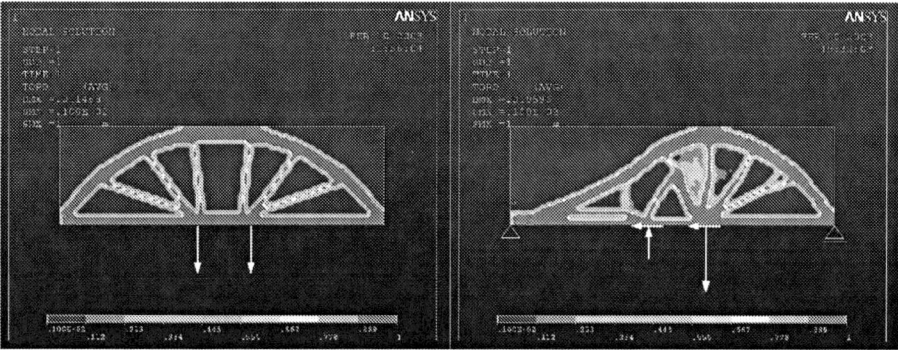

Figure 3.35.   Optimization results using commercial software and single load cases.

The same topology found by the original designer for the six panel truss was also found by the IGDT. In addition several others, including lighter topologies, were also found with the same or similar complexity (the first rows of Figures 3.31. and 3.32.). Figure 3.36. shows the IGDT geometry for the same topology as used by the designer in Figure 3.34. The IGDT chosen geometry is 28% lighter.

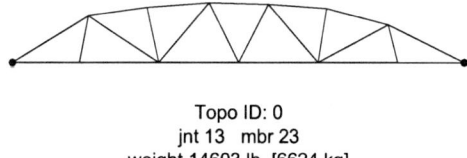

Topo ID: 0
jnt 13   mbr 23
weight 14603 lb [6624 kg]

Figure 3.36.   The IGDT geometry for the same topology as shown in Figure 3.34.

3.4.4   Conclusions

This example provides the opportunity to compare the results obtained using the IGDT with those found using commercial software. The IGDT provides advantages in two ways. First, as pointed out with respect to the other examples, by offering a choice of solutions, the IGDT goes much further toward intelligent interaction with the designer. Issues of complexity vs. weight reduction can be better explored when viewing a range of solutions such as that shown in Figures 3.31. or 3.32. In addition, the IGDT was able to give a more complete analysis in this case, by including the multiple variables of moving loads and the non-linear parameter of buckling. Thus, it is not surprising that the IGDT was able to offer a substantial improvement over the design found using commercial software.

3.5 Interactive Design

The preceding examples were all run using the automatic mode of the IGDT. This was primarily because they were all run by the author, and in order to avoid the impression of a 'rigged game' the selection was left to the program. In addition it is difficult to record and describe the design process that takes place when using the interactive mode. One sees the results, but the decisions made in selection remain in the designer's head, and may not even be conscious or expressible.

What is shown in this example are the results of a class assignment made to a group of architecture students. The class had visited a local wrought iron bridge and subsequently used it as a subject in learning finite element analysis techniques. The students were then asked to redesign the bridge for a standard moving truck and lane loading (HS 20). The desire was first to find a more weight efficient solution, but also to find a form that the designer felt was aesthetically pleasing and suited to the environment.

3.5.1 Problem Description and Setup

The Foster wrought iron bridge is shown in Figure 3.37. The design was patented in 1876 by the Wrought Iron Bridge Co. of Canton Ohio. It is a simple one lane Pratt through bridge design with all truss work above the deck. The 120 ft [36 m] span is divided into 8 panels.

Figure 3.37. The Foster Bridge in Ann Arbor, Michigan which was used as a case study and pattern for the student design exercise.

3.5.1.1 Problem Parameters and Geometry

Dimensions of the Foster Bridge truss are given in Figure 3.38. This was used as a progenitor truss for all above deck trials. The students were given the freedom to either limit the truss to above the deck or to allow members below the deck as well. Due to time constraints, the use to the IGDT was limited to one session of about 30 minutes. From that session each student group selected a truss which was then developed in 3D and finished by sizing all members using a commercial FEA package, STAAD.Pro 2004. STAAD used tubular steel sections and the requirements of the ASD-ASCI steel code to size all members and determine the total weights shown for each design.

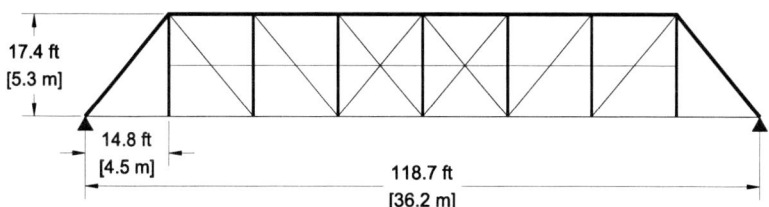

Figure 3.38. Dimensions of the original Foster Bridge.

3.5.1.2 Loading Cases

The loading was based on a standard AASHTO HS 20 moving truck and lane load. The axle load on the two axle truck was 32 kips [14.2 kN] and the lane load was 640 lb/ft [9.4 kN/m]. Since axle spacing for HS 20 can be between 14 ft. and 30 ft. [4.3 – 9.1 m], 14.8 ft [4.5 m] was chosen to match the panel spacing. Figure 3.39. shows the combinations of lane and moving load as they were applied to the truss. Self weight was left out of the IGDT analysis in order to allow faster run times. The self weight was, however, included in the STAAD analysis.

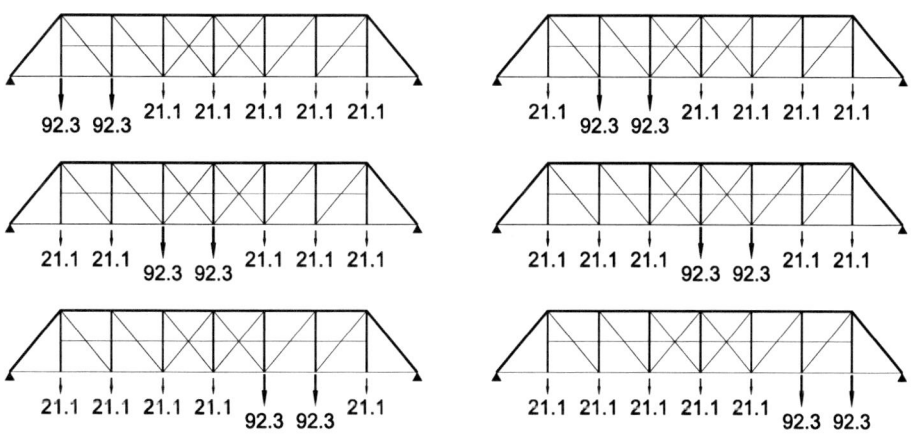

Figure 3.39. The combination of the six moving truck loads with the lane load. All values in kN.

3.5.2 Use of the IGDT

Due to time constraints the students were only able to use the IGDT in one session to generate forms. The sessions ran about 30 minutes which allowed for between 4 and 6 interactive cycles. Each group made the run independently and without knowledge of the work of the other groups.

3.5.2.1 Results of Student Designs

Figures 3.40. through 3.43. show the results of four of the groups. There were seven groups altogether and none of them chose the same design topology. Most designs selected tended to be in the same weight range as the original design, with some heavier and some lighter. Finding a lighter solution was an obvious objective, but some students treated weight more as a secondary objective compared to some formal requirement. For example the bridge shown in Figure 3.41. was a deliberate attempt to find a solution that included trusswork above and below the deck.

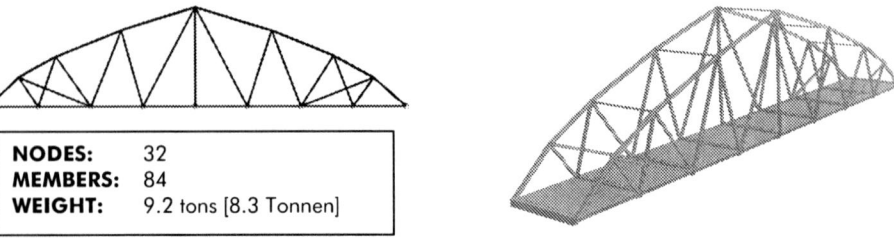

NODES: 32
MEMBERS: 84
WEIGHT: 9.2 tons [8.3 Tonnen]

Figure 3.40. Design by Yujin Kim, Jaewon Song and Jaewong Lee.

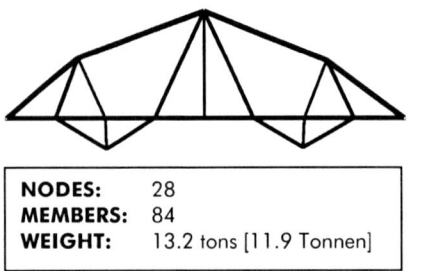

NODES: 28
MEMBERS: 84
WEIGHT: 13.2 tons [11.9 Tonnen]

Figure 3.41. Design by Robert Walsh.

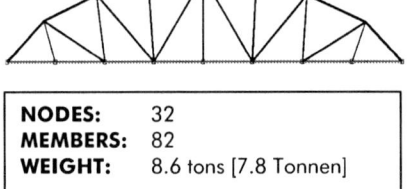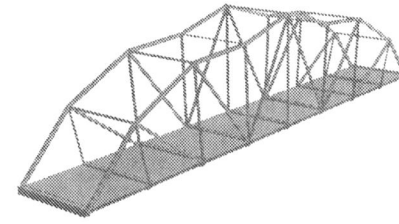

NODES: 32
MEMBERS: 82
WEIGHT: 8.6 tons [7.8 Tonnen]

Figure 3.42. Design by Xuezhen Chen and Jin Jeon.

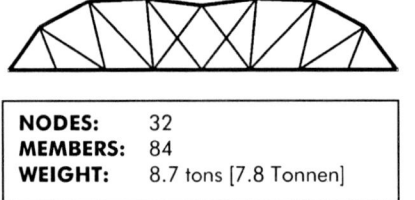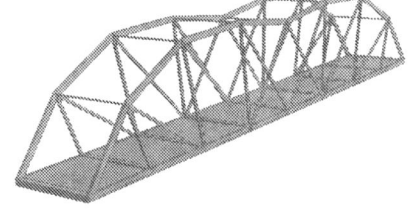

NODES: 32
MEMBERS: 84
WEIGHT: 8.7 tons [7.8 Tonnen]

Figure 3.43. Design by Lauren Bostic, William Marquez, Jennifer Siegel, and Faye Whittemore.

Although this design exercise was not configured to measure user reaction to the software interface, the students seemed to have no trouble with viewing the images on the screen or following the logic of making selections. The short one time session was not really adequate for the users to offer suggestions on the interface. It was likely the first time many of them had run a program in a Unix environment, and so they were understandably reluctant comment much on the interface. In addition, with the author of the program, who was at the same time their instructor for the class, watching the procedure, any comments they may have made were certainly less than completely candid. None the less, the general sense of the trial was that the program preformed as intended and the users were able to gain a definite insight and better understanding of the possible solutions.

3.5.3 Comparison of Results

Figure 3.44. shows the original Foster Bridge geometry redesigned using the STAAD member selection as was done with the student models so that the weight could be reasonably compared. The Foster design was much more detailed with almost double the number of members (actually elements from the FEA) as compared with the student designs. It also contains several tension-only member types that allow reduced size since the neither resist buckling or meet compression slenderness ratio requirements.

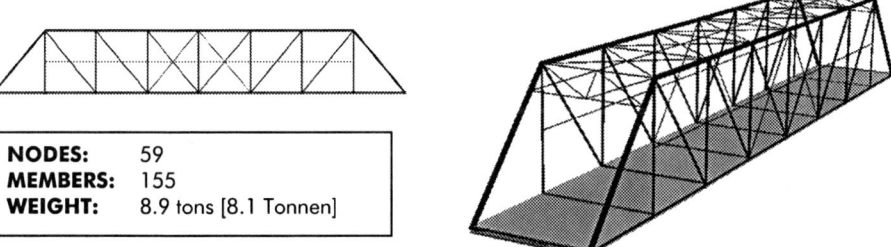

NODES: 59
MEMBERS: 155
WEIGHT: 8.9 tons [8.1 Tonnen]

Figure 3.44.   The original Foster Bridge from 1876.

The most obvious difference in the Foster design as compared with any of the other designs is the flat top which resulted in highly repetitive member lengths. The Wrought Iron Bridge Company originally sold these bridges through a catalog and mass produced them at the factory in Canton, Ohio. Not surprisingly, the architecture students were not particularly interested in issues of mass production and economy of repetition. It might also be considered that these are students, not experienced designers. Nevertheless, some of the designs did have merit. Both 4.42. and 4.43. are more weight efficient with far fewer members and fewer connections. In both cases the chosen form is also visually successful. The designer of the bridge shown in Figure 3.42. was specifically interested in marking the center axis of symmetry. The design group in Figure 3.43. took almost the opposite tact by dipping the center and creating a wave form.

What is perhaps most interesting is that no two groups followed the same path. In that respect the IGDT was very successful. Had the groups tried the same exercise but using a traditional optimization program which would yield one solution, it is rather certain that their designs would have been much more similar if not identical. Also, special features such as the peak at the center or the wave form would have been difficult to find.

3.5.4 Conclusions

Although this example is perhaps the least defined of the set presented, it is the only example which involved users other than the author. It is included primarily for that reason; to demonstrate that the IGDT can in fact be used by designers in exploring form. In this regard, the IGDT did just what it was intended to do. It exposed a variety of reasonable solutions to the designers, and allowed them to find the expression they were seeking in the context of structural efficiency.

Figure 3.45. Rendering of the bridge designed by student group: Lauren Bostic, William Marquez, Jennifer Siegel, and Faye Whittemore.

4 Conclusion

4.1 Summary

4.1.1 Aspects of Design

4.1.1.1 Producing Design

The term "design" has meaning that is context dependent, and discipline relative. Both architects and engineers see design as a complex, multi-phased process. The IGDT centers on the early phases of design development. As such, it is primarily concerned with issues of creativity and form exploration, rather than detail optimization and refinement. Creativity and form exploration is an area which has traditionally been hard to effectively support with computer aided tools. Nonetheless, decisions made in the early design phases have great impact on the relative success of a project. In projects where architectural form is closely tied to an understanding of the appropriate structural form, there has always been a need for tools which can be effectively used to explore the possible design space.

4.1.1.2 Aiding Design

Design tools are distinguished in this work from analysis tools in their ability to discover new forms or solutions, rather than being limited to the analysis of an existing design. Design tools are necessarily exploratory in nature. Whether model based, graphic or computer driven, good design tools stimulate the designer's own creativity in exploring the solution space. This means offering multiple solutions for the designer's consideration, rather than a single 'optimized' solution, which can be, in its own way, a detriment to creativity by causing fixation on that single, offered solution. The tool described in this work, the IGDT, is a form exploration tool that is able to reveal to the designer, arrays of 'pretty good' solutions. As a tool for exploration, it allows the user to direct its path through the design space in a way that conforms to the user's often less than articulate concept of the appropriate form vocabulary of a given project. The use of the tool is intended to be both stimulating to the designer's own creative instinct as well as supportive in revealing forms which are structurally suitable to the specific design conditions. In this way, the IGDT defines a new class of design tool that is particularly appropriate to the early phases of design development.

4.1.2 Aspects of GAs and the IGDT

4.1.2.1 GAs as Search Algorithms

Genetic Algorithms (GAs) are a class of evolutionary computation that are used as search engines in problems which are either ill structured or contain multiple or often conflicting objective parameters. In the last three decades much effort has been expended in the development of mechanisms and applications of GAs. It has been documented that although they will almost always be less efficient and less apt to find the very peak optimal solution as defined by the fitness function, they are much more robust and flexible in application. This work has found that to be true. In addition GAs have the capability of allowing user-interactive definition of the fitness function. User supplied fitness functions have been a fascination of artists for many years. Examples of work by Latham (Latham, et al., 1990), or Dawkins' classic "Biomorphs" (Dawkins, 1986) are clear examples. More recently there has been a realization of the possibilities offered by user-interactive GAs in solving engineering design problems (Boschetti, 2001).

4.1.2.2 Interactive Exploration

This work illustrates the use of GAs as aids in the exploration of possible alternate solutions within a design space. It was found that the use of the populations of solutions inherent to GAs, combined with the potential for user interaction, makes GAs ideal as tools for aiding design, particularly in the early stages. Section 1.2 discusses the nature of design, and what activities or methods can be employed to successfully aid design. Section 1.3.4 shows specifically how the IGDT contains many of these features.

It was determined that two features are key to the success of a design tool:

- exploration
- interaction

Exploration is the search of the design space for a pool of solutions from which selection can take place. It is important that the tool not limit the selection to one solution. It is not possible to select from a set of one. Only acceptance or rejection is possible in such an instance. This is not aid, but coercion. Although with traditional (non-computer based) design aids, it was more common to explore several different directions, it has been shown that the use of computers tends to promote the continued development of a single design direction (Goel, 1992). This is stifling to creativity, and rapidly leads to design fixation. By always presenting the user with a selection set of alternative solutions, the IGDT overcomes the danger of fixation in early phases of design, and is able to function as a true exploration tool.

Interaction is the other critical ingredient for a design development tool. Design is an engaging activity. It is more than the simple selection of a solution. It involves the definition of the problem itself, and therefore, suffers under rigidly defined and limiting objective functions. The IGDT is successful in engaging the designer by allowing user-interactive definition of the fitness function. In this way the designer's own expertise and experience are brought into play. The tool still requires the craftsman, which is both satisfying for the designer as well as enriching for the designed product.

4.2 Results and Recommendations

The discussion of results and the conclusions drawn from each example are presented at the end of each example in Section 3. What is discussed in this section is the overall characteristics of the tool rather than specific instances of its performance.

4.2.1 Applications

As is the case with most optimization procedures, the time investment usually makes the application uneconomical for routine use. The time schedule constraints placed on typical architectural projects, simply can not allow a detailed exploration of multiple directions. This is why architectural researchers commonly pursue first their research topic, and then find an opportunity to apply the results. Nonetheless, some projects make better candidates for detailed structural consideration than others. One example would be a project in which some structural element plays a particularly significant roll. Another example would be a project in which the repetition of some element makes it worthwhile to give that element particularly detailed consideration.

It is to the advantage of the IGDT concept that the problem input is no more complex than low end FEA program (e.g., STAAD) which is routinely used in offices. There are numerous examples on the market where FEA programs have been coupled with CAD programs and a rather intuitive input interface, allowing even the beginning student ready access. As it now stands, the major limitation to the IGDT concept is the computational intensity inherent to GAs. Although this translates into long run times with today's technology, the time may soon arrive when computing speeds increase sufficiently to make GA applications more accessible.

4.2.2 Current Limitations

4.2.2.1 Hardware - Software

The coding for the IGDT was executed in ANSI standard C. The coding in ANSI C should greatly facilitate porting to other platforms. However, as mentioned above, the computational demand is fairly high. The examples in Section 3. were all run in parallel on a distributed network with 30 hosts using the Parallel Virtual Machine (PVM) message-passing software (Geist, et al., 1994). Even with this capacity, problems with multiple load cases like the arch truss example, Section 3.2., or where self weight is considered, take several hours or overnight to run. Although run time is hardware dependent, since one of the major drawbacks of GAs and the IGDT is long run times, it is useful to chart final run times for the examples used in Section 3. Table 4.2. shows run times (median of 3 runs) for each example problem. In forming this chart all examples were re-run with the same hardware configuration and, as far as possible, the same parameter configuration to aid in making an assessment. The parameters shown in Table 4.1. were used consistently in all examples. The complete list of parameters are listed in Appendix A in the #define section of truss.h.

Some of the other #define parameters were set differently on different examples. The analysis routine on one example (option 1 of example 4.3.) was set not to include buckling. Otherwise the same steel pipe design routine (AISC-ASD) including buckling was used for all compression members. All tension members were designed as steel rods, likewise based on the criteria of AISC-ASD.

For university settings with large pools of machines that can readily be networked, distributed computing is possible. In fact, in such settings, substantially more hosts can often be found without too much trouble. But for a medium to small design firm this would certainly represent a limitation.

Value	#define	Definition
2	START_UP	Use progenitor to initiate first topo generation
run_x_cyc	RUN_GA	Run geometry GA for a set (x=3) number of cycles
100	GENERATIONS	Maximum number of generation in the geometry GA
4	CYCLES	Number of cycles run in the geometry GA
50	POP	Size of the parent population in the geometry GA
select_elite	SELECT_TOP	Topology selection method (elitist=choose best)
2	TOPO_OP	Run a set number of cycles in automatic mode
40	TP_GEN_NR	Maximum generation in one topology cycle
7	TP_CYC_NR	Set number of cycles run in the topology GA
15	CYC_SAME	No. of topos in a population defining convergence
20	TOPO_NR	Number of parents in a topology generation
SPAN/30	MELD_LIMIT	Joints found closer that this are joined to one joint
SPAN/30	CUT_LIMIT	Joints closer to members are joined to member
SPAN/120	DEFLECT_LIM	Solutions beyond this limit are penalized

Table 4.1. The more important #define parameters that were held constant for the timed trials.

Example	Joints Start	Joints End	Fixed Joints	Max. Joints Allowed	Load Cases	Self Weight	Buckling	Time (hours)
Pony Bridge	17	18	5	27	1	Yes	Yes	1.5
Deck Bridge	17	11	5	27	1	Yes	Yes	0.44
Lenticular Bridge	30	16	5	36	1	Yes	Yes	1.8
Arch Truss	15	15	9	27	6	No	Yes	3.8
Cantilever 1	11	11	3	27	1	No	No	1.0
Cantilever 2	11	9	3	27	1	No	Yes	1.8
6 Panel Gantry	20	28	7	30	10	No	Yes	12.6
7 Panel Gantry	24	33	8	35	12	No	Yes	39.7
Interactive	16	~16	9	30	6	No	Yes	~0.6*

Table 4.2. Comparison of run times (median of 3 runs) for the examples from Section 3.
* on the interactive trial, each group ran about 4 topology cycles at 10 min. each.

4.2.2.2 Personnel Requirements

The average architectural engineering personnel should be able to use the IGDT with out too much trouble. In its present form, documentation is lacking, and the reliance on the direct manipulation of #define parameters in the header file is not very user friendly. But the actual input for the IGDT is no more complex than the average FEA program. In fact the one input file that is required was patterned after the data files used by SAP IV (Bathe, 1974), which had a fairly typical FEA input format. What is lacking for general usage is a modern graphic input interface. I am currently working on a web based graphic interface. Since the actual parallel cluster is usually accessed remotely anyway, having the interface readable over the internet would allow easy access via any web browser worldwide.

The final example in Section 3.5, makes use of students at the University of Michigan to operate the program. The trial was made in the context of a structural framing class that had used the FEA software, STAAD-Pro, earlier in the semester. With less than an hour of instruction the students had no trouble in using the IGDT to explore alternative topologies for a local truss bridge. What they were not required to do themselves was to set the #define parameters before execution. That would have to be made more accessible before general users could define parameters themselves.

However, the IGDT itself functioned as expected and provided the users with useful input to the design process. Nonetheless, it remains in principle a search tool that requires an experienced researcher. Without a craftsman to guide its use, the results can not be any better than what is reached in the automatic mode, and a true exploration of the design space would not be achieved.

4.2.2.3 Time

As mentioned above, the biggest limitation is certainly the time required for a run. The current parallelization was coded with the premise that the number of hosts available would be on the order of the topology population size. For the examples run in Section 3 this was usually taken at 20. As of 2006, the Hydra Linux cluster build for the IGDT contains 100 nodes. Since the size of the topology population is tied to the number of members in a single individual solution, this larger cluster will allow the practical exploration of larger 3D type systems. Unfortunately, there would not continue to be the same scalable speedup as the cluster size exceeded the population size. To see further gains from the parallel network at that level would require some additional coding to parallelize the geometry GA. On the other hand, if CPU speeds continue to increase, as seems likely, then eventually these long run times will become less of a problem.

4.2.3 Further Development

4.2.3.1 Team Design

Although the popular image of a designer is the single genius working alone, the much more common scenario is a team of professionals working together. There is certainly much to be gained in the cross fertilization of ideas. With the IGDT already running in a distributed environment, it would be easy to make the results accessible to a project team. In this way several designers could participate in the interactive exchange with the IGDT. In exploring more complex geometries this may offer the advantage of pooling the ideas of several designers. Some architects have already suggested genetically encoding the style of great designers so as to be able to continue to clone their work after their death (Section 1.3.3.7). I would rather propose to build libraries of good solutions, and give future architects a means of accessing and further developing these designs. The fundamental premise of the IGDT concept is after all, to aid and support the human designer, not to clone and replace him.

4.2.3.2 Code Development

The IGDT developed for this dissertation was coded as an entirely independent application. That is, with the exception of PVM which manages the parallel message passing and AutoCAD which is used to plot the output, no use is made of other commercial modules. This was done deliberately to maintain sufficient control over the design of the process, as well as to increase the computational speed of the program.

After having worked through the examples in Section 3, a few additional routines could be suggested that would enhance the performance of the IGDT. These can be categorized according to the level of search:

- geometry
- topology

Geometry level improvements center on the development of additional element capabilities. Currently there is only one element - the truss element with two degrees of freedom. A frame element with three degrees of freedom, would considerably expand the range of possible geometries. Ideally, a six degree of freedom 3D beam element would expand the range of explorable geometries to include most architectural framed structures. However, the computational cost would be considerable particularly as the overall size of the systems would also increase. Remaining with reticulated structures (trusses and frames), other element enhancements would include: tension only elements (cables), or compression only elements, or pre-stressed elements. Further enhancements could be developed in the area of loadings. Linear or surface type loadings would be useful when beam elements are added. The difficulty with area type loadings is that since not only the geometry but also the topology is always changing, the loading conditions become a little difficult to define. It can likely be done, but also requires more computational capacity.

As mentioned above, one strategy to increase computational capacity is to simply add more machines to the parallel cluster. But since jobs are currently partitioned based on the topology population, and since the cluster is already currently about that size, this would not produced a scaled increase in performance. In addition, the job partitioning would have to be shifted to the geometry GA as the basic unit. This could allow the efficient usage of clusters as large as the simultaneous number of geometry optimizations occurring in one topology cycle. For the examples of the scale run in Section 3 this would be 50 geometry GAs x 20 topology GAs = 1000 possible parallel operations. So partitioning the parallel problem based on the geometry GA rather than the topology GA would make the code scaleable up to about 1000 nodes rather than 20 as is currently the case. Although 1000 is quite a few nodes, numbers like 200 are currently not so uncommon (the IGDT currently runs on a 100 node Linux cluster), and lager clusters currently exist such as the 2000 node Linux cluster at SUNY Buffalo.

Topology level improvements are not so obviously needed as geometry. The inclusion of 3D is the most interesting thing that could be enhanced. It is also conceivable that continuous surface structures could be investigated using the IGDT. This would of course require the inclusion of the appropriate finite elements, and more importantly specific routines to handle breeding and mutation.

As it currently is coded, the FEA is integral with the GA coding. Although they involve separate routines, they access common code structures without translation or even copying of data. All structure manipulation is through c pointers. This allows the code to run very quickly. Although it is tempting to harness an existing FEA package in order to gain the added capability, there would be a rather high price to pay in decreased run time efficiency. For a geometry population of 50 run through 4 cycles, each with an average generation number of 100 (the configuration used in the examples), during one optimization process the GA calls the analysis routines on the order of 10,000 times (50/2 x 100 x 4). With a topology population of 20, averaging 7 generations to convergence, and run through 7 cycles, the number of geometries that would have to be optimized would be about 1000 (20 x 7 x 7). So the total number of geometries analyzed for one topology optimization is the product of the two, i.e., about 10 million (10,000 x 1,000). If each analysis takes only 0.1 of a second, this yields 278 hours or 11.5 days. This is assuming a small truss like the one used in the first example of Section 3. The necessity of parallel computation as well as efficient code is obvious.

Because the actual number of computations depends on how quickly both geometry and topology populations converge, the number of computations per run will vary. The times shown in Table 4.2. reflect the parameter dimensions described above. The times are further effected by including self weight (2-4 further iterations), additional load cases, and buckling calculations. The overhead of passing data, loading and unloading a commercial FEA module on the order of 10 million times would certainly be prohibitively large. For example, on a 1.3 GHz P-IV, running STAAD.Pro under Windows 2000, with the program already opened, it takes about 2 seconds to load a small data file for the bridge truss in the first example in Section 3. It takes another 5 seconds to run it and another 5 seconds if members are sized. Assuming 10 seconds for the whole operation, at 10,000,000 times that would take the P-IV 1150 days or 3.2 years. This may be a good argument not to use MS-Windows as a platform, but the point is that a lot of computational overhead exists in any commercial package.

In the current IGDT application the FEA data is all contained in four global structures, which only need to be referenced by pointers. The data structures can thus be shared and accessed by both the FEA and GA routines. This is of course much faster than actually moving data. Perhaps a compromise solution would be to integrate an existing FEA package, such as SAP IV (Bathe, 1974) which has public domain code, into the IGDT. This would save some time in re-coding, but in the end might be less flexible, and still include unneeded code.

4.2.3.3 Library Development

In the coding of the IGDT, use was made of only the standard C libraries, plus the special PVM library. There are 177 sub-routines. It would make future development somewhat simpler to extract common code, and form an IGDT library. A second library could perhaps gather code used in graphics manipulation for input and output files.

4.2.3.4 Platform

Although all development to this point has been in UNIX (or Linux), the ANSI C coding as well as PVM message passing is open to all major platforms. An advantage of using PVM is that is has the ability to combine different host platforms into a parallel distributed network. It is only necessary to be able to compile the codes (both PVM and IGDT) on each platform.

4.2.3.5 Parallelization

Because GAs deal with populations, and each individual in the population must be assigned a fitness through some type of analysis, it is quite easy to partition the problem for parallel processing based on the individuals in the population. In the case of the IGDT, the population of the higher level topology GA was used as a basis in partitioning the code for parallel processing. In this form the IGDT operated very satisfactorily for the scale of the test problems and cluster size. The topology population was usually taken at 20 parents with 20 children, and the size of the cluster used was similar (starting at 10 and later at 30 nodes).

But in addition to the obvious benefit of increased speed, it was found that the program was much more robust when run in parallel than when run entirely as one executable on one host. With the heavy use of random input, unanticipated combinations of data and operation do occur, and can cause the system to fault. If this happens to one of many hosts, there is no great problem. The non-responding host can be detected and reinitialized without hindering the overall operation of the IGDT. This capability is, in fact, coded into the IGDT. Naturally, if the program is running on only one host, a crash is more critical.

From the discussion of computational demand described above, it is also obvious that any future expansion of the capabilities of the IGDT would have to rely heavily on parallel processing. On the level of single workstations, one could expect to generate solutions, either based on a single topology (geometry optimization), or limited topology mutation and breeding (small populations). But for larger topology explorations with more complex structures, a cluster is more suited.

4.3 Closing Remarks

The significance of the IGDT as a structural design tool can be summarized as follows:

- DESIGN
  - Aids exploration by exposing a set of 'pretty good' solutions
    - choice requires more than one solution
    - 'good' solutions are sub-optimal (local optima) peaks
  - Allows user interaction to steer the results
    - allows user selection
    - allows non-coded criteria (e.g., aesthetics)
- IMPLEMENTATION
  - GAs are inherently design oriented
    - offer selection
    - contain some surprises
  - Visual assessment is critical
    - interactive on screen
    - workable output in CAD format
  - Computational speed is critical
    - Avoid ground structures
    - Use parallel clusters

None of these points individually are anything new. Numerous examples and support material have been provided in this work which supports these points individually. GAs have been used in structural optimization, but only in ways which mimic traditional optimization techniques which produce single solutions. Design methods offer numerous ways to enhance exploration, but they are never coupled with structural design tools.

The novel aspect of the IGDT concept lies more in the realization and collection of these attributes into one design tool. It can thereby be rightly termed a new class of design aid, an "Intelligent Genetic Design Tool."

Figure 4.1. University of Michigan architecture students using the IGDT in the interactive mode. (photo by Prof. Mojtaba Navvab)

5 Reference List

Abel, John F. "Computer-Aided Design and Numerical Methods", in: Abel, J., Astudillo, R., and Srivastava (eds.) *Current and Emerging Technologies of Shell and Spatial Structures - Proceedings*, IASS/CEDEX, Madrid, 1997. pp. 189-200.

Adams, James L. *Conceptual blockbusting: a guide to better ideas*. Freeman, San Francisco, 1974.

Addis, William. *Structural Engineering: the nature of theory and design*. Ellis Horwood, London, 1990.

Adeli, H. and Cheng, N. "Concurrent Genetic Algorithms for Optimization of Large Structures", in: *Journal of Aerospace Engineering*. Volume 6, Number 4. p. 315. October, 1993.

Alexander, Christopher. *Notes on the synthesis of form*. Harvard University Press, Cambridge, Mass, 1967.

Anonymous "Theory and Practice", in: *The Practical Mechanic and Engineer's Magazine*, Oct. 1842. p. 1.

Antoniades, A. C. *Poetics of Architecture: Theory of Design*. Van Nostrand Reinhold, New York, 1990.

Archer, B. L. "An overview of the structure of the design process", in: Moore, G. T. (ed.) *Emerging Methods in Environmental Design and Planning*. MIT Press, Cambridge, Massachusetts, 1970.

Asimow, Morris. *Introduction to Design*. Prentice-Hall, Englewood Cliffs, New Jersey, 1962.

Austin, James H. *Chase, change and creativity: the lucky art of novelty*. Columbia Univ. Press, New York, 1978.

Bäck; Hoffmeister and Schwefel. "A Survey of Evolution Strategies", in: *Proceedings of the Fourth International Conference on Genetic Algorithms*. Morgan Kaufmann Publ., San Mateo, California, 1992.

Barron, F. "Putting creativity to work", in: Sternberg, R. J. (ed.) *The Nature of Creativity*. Cambridge University Press, Cambridge, England, 1988.

Bathe, Klaus-Jürgen; Wilson, Edward; Peterson, Fred. *SAP IV A Structural Analysis Program for Static and Dynamic Response of Linear Systems*. A Report to the National Science Foundation: EERC 73-11, Univ. of California, Berkeley, 1973 (rev. 1974).

Bendsøe, M. and Kikuchi, N. "Topology and Layout Optimization of Discrete and Continuum Structures", in: Kamat, Manohar P. (ed.) *Structural Optimization: Status and Promise*. Volume 150 Progress in Astronautics and Aeronautics. American Inst. of Aeronautics and Astronautics, 1993. pp. 517-547.

Boden, Margaret A. What is Creativity?, in: Boden, M. A. (ed.), *Dimensions of Creativity*. The MIT Press. Cambridge, Massachusetts. pb. ed. 1996 (1st ed. 1994).

Boschetti, F. and Moresi, Louis. "Interactive Inversion in Geosciences", in: *Geophysics*. Volume 64. pp. 1226-1235, 2001.

Bouzy, C. and Abel, J. F. "A Two Step Procedure for Discrete Minimization of Truss Weight", in: *Structural Optimization 9*. pp. 128-131, 1995.

Broadbent, G. *Design in Architecture*. John Wiley & Sons, 1973

Brooks, R. A. "Fast, Cheap and out of Control: A Robot Invasion of the Solar System", in: *Journal of The British Interplanetary Society, Vol. 42*, 1989.

Burkhardt, Berthold. *IL1: Minimal Nets*. Karl Krämer Verlag, Stuttgart, 1969.

Cai, Jianbo. *Diskrete Optimierung dynamisch belasteter Tragwerke mit sequentiellen und parallelen Evolutionsstrategien*. Essen Univ., Diss., 1995.

Candela, F. Autographed quotation from a lecture at Ove Arup & Partners. c.1950.

Cheng, Nai-Tsang. *Integrated genetic algorithms for optimization of structures*. Thesis (M. S.)--Ohio State University, 1992.

Coello Coello, C.A. "An Introduction to MOEAs and Their Applications", in: Coello Coello & Lamont (eds) *Applications of Multi-Objective Evolutionary Algorithms*. Advances in Natural Computation, Vol.1. pp. 1-28. World Scientific, 2004.

Coello Coello, C.A. "Discrete Optimization of Trusses Using Genetic Algorithms", in: *Expert Systems Application & Artificial Intelligence - EXPERSYS - 94*. p. 331. IITT International, 1994.

Coello Coello, C.A.; Rudnick, M. and Christiansen, A. D. "Using Genetic Algorithms for Optimal Design of Trusses", in: *Proceedings. Sixth International Conference on Tools with Artificial Intelligence* (Cat. No. 95CH35727). p. xxiii+842, 88-94. IEEE Comput. Soc. Press, Los Alamitos, CA, USA, 1994.

Colbron, B., Gero, J. S., Purcell, T. and Williams, P. "The role of design discipline and pictorial information in fixation effects in design problem solving", in: *CogSci '93*, 1993. pp. 76-78.

Conybeare, H. "On the Principles and Practice of Civil Engineering", in: *Civil Engineer & Architect's Journal, Vol. 21*, 1858.

Corbusier, L. *The Chapel at Ronchamp*. Frederick A. Praeger, New York, 1958.

Coyne, R. D.; Roseman, M. A.; Radford, A. D.; Balachandran, M. and Gero, J. S. *Knowledge-Based Design Systems*. Addison-Wesley, Reading, Mass, 1990.

Crovitz, Herbert F. *Galton's Walk: Methods for the Analysis of Thinking, Intelligence, and Creativity*. Harper & Row, New York, 1970.

Crowe, N. and Lasweau, P. *Visual Notes for Architects and Designers*. John Wiley & Sons, 1984.

Cryer, John N. "Design Team Agreements 3.43", in: Haviland, David (ed.) *The Architect's Handbook of Professional Practice, Vol. 2 The Project*. AIA Press, Washington, D.C. 1994.

Dawkins, Richard. *The Blind Watchmaker*. Norton, W.W. & Co., London, 1986.

de Bono, Edward. *Lateral Thinking for Management: A Handbook for Creativity*. American Management Association, New York, 1971.

De Jong, K. A. *An Analysis of the Behavior of a Class of Genetic Adaptive Systems.* University of Michigan, Diss, 1975.

de Kleer, J. and Brown, J. S. "Qualitative physics based on confluences", in: *Artificial Intelligence* Vol. 24,1984.

Deb, K. and Gulati, S. *Design of truss-structures for minimum weight using genetic algorithms.* KanGAL Report No. 99001. Kanpur: Kanpur Genetic Algorithms Laboratory, Department of Mechanical Engineering, Indian Institute of Technology Kanpur, Kanpur 208016, India. 1999.

Do, E. and Gross, M. D. "Supporting Creative Architectural Design with Visual References", in: J. Gero et al. (ed.), *3rd International Conference on Computational Model of Creative Design (HI '95).* Australia, 1995.

Dunker, K. "On problem solving", in: *Psychological Monographs*, 1945.

Duysinx, P.; Zhang, W.; Fleury, C.; Nguyen, V.; and Haubruge, S. "A New Seperable Approximation Scheme for Topological Problems and Optimization Problems Characterized by a Large Number of Design Variables", in: Olhoff, N. and Rozvany, G. (ed.) *WCSMO-1: Proceedings of the First World Congress of Structural and Multidisciplinary Optimization.* 28 May - 2 June 1995, Goslar, Germany. Pergamon, 1995. pp. 1-8.

Dym, Clive L. and Levitt, Raymond E. *Knowledge-Based Systems in Engineering.* McGraw-Hill, New York, 1991.

Edwards, Betty. *Drawing on the right side of the brain: a course in enhancing creativity and artistic confidence.* Tarcher, Los Angeles, CA, 1979.

Eshelman, Larry. "Productive Recombination and Propagating and Preserving Schemata", in: *Foundations of Genetic Algorithms 3.* Morgan Kaufmann Publ., Inc. San Francisco, 1995.

Eshelman, Larry. "The CHC Adaptive Search Algorithm: How to Have Safe Search When Engaging in Nontraditional Genetic Recombination", in: *Foundations of Genetic Algorithms.* Morgan Kaufmann Publ., San Mateo, California. 1991.

Fabrycky, W. J. and Blanchard, B. S. *Life-Cycle Cost and Economic Analysis.* Prentice-Hall, Englewood Cliffs, New Jersey, 1991.

Frazer, J. *An Evolutionary Architecture.* Architectural Association, London, 1995.

Führer, W. "Rechnerunterstützter Tragwerksentwurf von Stab- und Flächentragwerken", in: *CAAD - Fortschritte bei uns und unseren Nachbarn: Rechnerunterstützte Informationsverarbeitung in der Architektur.* Technische Universität Berlin, Berlin, 1991. pp. 105-117.

Galante, M. "Genetic Algorithms as an Approach to Optimize Real-world Trusses", in: *International Journal for Numerical Methods in Engineering.* Volume 39, Number 3. p.361, 1996.

Geist, ; Beguelin; Dongarra; Jiang; Manchek; Sunderam. *PVM: Parallel Virtual Machine. A Users' Guide and Tutorial for Networked Parallel Computing.* The MIT Press, Cambridge, Massachusetts, 1994.

Gero, J. S. "Creativity, Emergence and Evolution in Design: Concepts and Framework", in: *Knowledge-Based Systems.* 1997.

Gero, J. S. "Towards a model of exploration in computer-aided design", in J. S. Gero and E. Tyugu (eds.), *Formal Design Methods for CAD*. North-Holland, Amsterdam, 1994. pp. 315-336.

Gero, J. S. and Ding, L. Exploring style emergence in architectural designs, in Y-T. Liu, J-H. Tsou and J-H. Hou (eds.), CAADRIA'97, Hu's Publisher, Taipei, Taiwan, 1997. pp. 287-296.

Gero, J. S. and Jun, H. Visual semantic emergence to support creative design: A computational view, in J. S. Gero, M. L. Maher and F. Sudweeks (eds.), Preprints Computational Models of Creative Design, University of Sydney, 1995. pp. 87-117.

Goel, V. "'Ill-structured representations' for ill-structured problems", in: *Fourteenth Annual Conference of the Cognitive Science Society- Proceedings*, Bloomington, Ind., 1992. pp. 844-849.

Goldberg, David E. *Genetic algorithms in search, optimization, and machine learning*. Addison-Wesley, Reading, Mass, 1989a.

Goldberg, David E; Korb, Bradley; Deb, Kalyanmoy. Messy Genetic Algorithms: Motivation, Analysis, and First Results". In: *Complex Systems*, Vol 3, No 5 Oct 1989b.

Gordon, William J. J. *Synectics: the Development of Creative Capacity*. Harper & Row, New York, 1961.

Grierson, D. and Pak, W. "Discrete Optimal Design Using a Genetic Algorithm", in: *Topology Design of Structures - NATO Advanced Research Workshop*. Kluwer Academic Publ., 1993.

Gross, M. D. "Recognizing and Interpreting Diagrams in Design", in: Catarci, T.; Costabile, M.; Levialdi, S.; Santucci, G. (eds.) *Advanced Visual Interfaces '94 (AVI '94)*. ACM Press, 1994.

Hajela, P and Lee, E. "Genetic Algorithms in Truss Topological Optimization", in: *International Journal of Solids and Structures*. Volume 32, Number 22. pp. 3341, 1995.

Hale, M. A. "An Open Computing Infrastructure that Facilitates Integrated Product and Process Development form a Decision-Based Perspective", Doctoral Dissertation, Georgia Institute of Technology, School of Aerospace Engineering, July, 1996.

Höfler, A. *Form Optimierung von Leichtbaufachwerken durch Einsatz einer Evolutionsstrategei*. Dr. Ing. Diss., Technische Universität Berlin, 1976.

Holland, John H. *Adaptation in Natural and Artificial Systems*. The Univ. of Mich. Press, 1975.

Isaken, S. and Trefflinger, D. J. *Creative Problem Solving: The basic course*. Bearly Limited, Buffalo, New York, 1985.

Johnson, S.; von Buelow, P. and Tripeny, P. "Linking Analyses and Architectural Data: Why It's Harder than We Thought" in Beesley, Cheng and Willismson (eds) *Fabrication: a conference examining the digital practice of architecture*, Proceedings of The Association for Computer-Aided Design in Architecture. Toronto, Canada, 2004.

Jun, H. and Gero, J. S. "Representation, Re-representation and Emergence in Collaborative Computer-aided Design", in: Maher, M. L., Gero, J. S. and Sudweeks, F. (eds.), *Preprints Formal Aspects of Collaborative Computer-Aided Design*, Key Centre of Design Computing, University of Sydney, Sydney, 1997. pp. 303-320

Kelly, Kevin. *Out of Control: The New Biology of Machines, Social Systems and the Economic World*. Addison-Wesley, Reading, Mass, 1995.

Kikuchi, N., Cheng, H.-C., and Ma, Z.-D. "Topological Design for Vibrating Structures", In: *Computer Methods in Applied Mechanics and Engineering*. Vol.121, 1995. pp.259-280.

Kirsch, Uri; "Layout Optimization Using Reduction and Expansion Processes", in: Olhoff, N. and Rozvany, G. (ed.) *WCSMO-1: Proceedings of the First World Congress of Structural and Multidisciplinary Optimization*. 28 May - 2 June 1995, Goslar, Germany. Pergamon, 1995. pp. 95 - 102.

Klarbring, A.; Petersson, J. and Rönnqvist, M "Truss Topology Optimization Involving Unilateral Contact - Numerical Results", in: Olhoff, N. and Rozvany, G. (eds.) *WCSMO-1: Proceedings of the First World Congress of Structural and Multidisciplinary Optimization*. 28 May - 2 June 1995, Goslar, Germany. Pergamon, 1995. pp. 129-134.

Knight, T. W. "Shape grammars: five questions", in: *Environment and Planning B: Planning and Design*, Vol. 26, 1999b.

Knight, T. W. "Shape grammars: six types", in: *Environment and Planning B: Planning and Design*, Vol. 26, 1999a.

Knight, T. W. "The generation of Hepplewhite-style chair-back designs", in: *Environment and Planning B: Planning and Design*, Vol. 7, 1980.

Knight, T. W. "Transformations of De Stijl art: the paintings of Georges Van Tongerloo and Fritz Glarner", in: *Environment and Planning B: Planning and Design*, Vol. 16, 1989.

Knight, T. W. "Transformations of the meander motif on Greek Geometric pottery", in: *Design and Computing*, Vol. 1, 1986.Kocvara, M. and Zowe, J.. "How to optimize mechanical structures simultaneously with respect to topology and geometry". In: N. Olhoff and G.I.N. Rozvany (eds.), *Structural and Multidisciplinary Optimization*. Elsevier Science, Oxford, 1995. pp. 135--140.

Koberg, Don and Bagnall, Jim. *The Universal Traveler - a Soft-Systems guide to: creativity, problem-solving, and the process of reaching goals*. William Kaufman, Los Altos, California. 1972.

Kolodner, *The Archie Project*. Georgia Institute of Technology. 1996.

Koning, H. and Eizenberg, J. "The language of the prairie: Frank Lloyd Wright's prairie houses", in: *Environment and Planning B: Planning and Design*, Vol. 8, 1981.

Koumousis, V. K. and Georgiou, P. G. "Genetic Algorithms in Discrete Optimization of Steel Truss Roofs", in: *Journal of Computing in Civil Engineering*. Volume 8, Number 3. p. 309, 1994.

Koza, John; Bennett, Forest; Andre, David; Keane, Martin. *Genetic Programming III: Darwinian Invention and Problem Solving*, Morgan Kaufmann Publishers, San Francisco. 1999.

Kuhn, Thomas S. *The structure of scientific revolutions*. Univ. of Chicago Press, Chicago, 1962.

Latham, William; Todd, Stephen and Owen, Mark. "Animating Abstract Forms", in: *Computer Graphics 90*. (Proceedings of the conference held in London, November 1990). Blenheim Online, London, 1990.

Leite and Topping. "Improved Genetic Operators for Structural Engineering Optimization". in: Topping, B.H.V. (ed.). Developments in Neural Networks and Evolutionary Computing for Civil and Structural Engineering. Civil-Comp Press, Edinburgh, UK, 1995.

Louis, S. J. and Fang Zhao. "Domain knowledge for genetic algorithms", in: *International Journal of Expert Systems Research and Applications*. Vol. 8, No. 3, p. 195-211. JAI Press, USA, 1995.

Luger, George F.; Stubblefield, William A. *Artificial intelligence and the design of expert systems*. Benjamin/Cummings Publ., Redwood City, Calif., 1989.

Manual of Steel Construction, Allowable Stress Design. ninth ed. American Institute of Steel Construction, Inc., Chicago, Illinois, 1989.

Martin, Harold C. *Introduction to Matrix Methods of Structural Analysis*. McGraw-Hill, New York, 1966.

Maxwell, J. C. *The Scientific Papers of James Clerk Maxwell*. Volume II. ed. by Niven. Librairie Scientifique J. Hermann. Paris. orig. 1890.

Mehr, A.F.; Azarm, S. " An Information-Theoretic Performance Metric for Quality Assessment of Multi-Objective Optimization Solution Sets". in: ASME Journal of Mechanical Design, Vol. 125(4). 2003. pp. 655-663.

Mitchell, A. G. M. "The Limits of Economy of Material in Frame-structures", in: *Philosophical Magazine and Journal of Science*. Volume VIII, Sixth Series, July - December 1904. Taylor and Francis, London, 1904.

Mitchell, M. *An introduction to genetic algorithms*. The MIT Press, Cambridge, Mass., 1996

Moneo, R. "On typology", in: *Oppositions*. Vol. 13, 1978.

Moro, Jose Luis. "About Showing Pictures and Being Unspecific" in: *Journal of the International Association for Shell and Spatial Structures*. Volume 37, Number 1. IASS, Madrid, Spain, 1996.

Nadin, Mihai. "Computational Design", in: *Formdiskurs: Zeitschrift für Design und Theorie*. Volume 2,1. 1/1997.

Nervi, P. L. *Structures*. F. W. Dodge Corporation, New York, 1956.

Newell, A. and Simon, H. *Human Problem Solving*. Prentice Hall, Englewood Cliffs, New Jersey, 1972.

Onions, C. T. *The Shorter Oxford English Dictionary on Historical Principles*, Oxford University Press, Oxford, 1968.

Orta-Rial, Belén. "Trusses Optimisation Under Multiple Variable Load" In: *Computing in Civil and Building Engineering: Proceedings of the Eighth International Conference.* Vol. 2, ASCE, Stanford, California, August 2000. pp. 1450-1457.

Osborn, A. F. *Applied Imagination: Principles and Procedures of Creative Thinking.* Scribner's, New York, 1963.

Parnes, S. *Sourcebook for Creative Problem Solving.* Creative Education Foundation Press, Buffalo, New York, 1992.

Peter, Boris. *Druckbeanspruchte Glasdachkonstruktionen.* Diplomarbeit (Thesis), Institut für Leichte Flächentragwerke, Universität Stuttgart, May 2000.

Pinker, S. *The Language Instinct*, Harper Collins Publishers, 1995.

Pippard, A. J. S. "Education and Theory", in: *The Engineer*, Centenary Number, 1956.

Plsek, Paul E. *Creativity, Innovation, and Quality.* Quality Press, Milwaukee, 1997.

Powell, D., Skolnick, M. "Using Genetic Algorithms in Engineering Design Optimization with Non-linear Constraints", in: *Proceedings on the Fifth International Conference on Genetic Algorithms.* Morgan Kaufmann Publ., San Mateo, California. 1993.

Prince, George M. *The Practice of Creativity: a Manual for Dynamic Group Problem Solving.* Harper & Row, New York, 1970.

Purcell, T. A. and Gero, J. S. "Design and other types of fixation", in: *Design Studies* 17(4): 1996. pp. 363-383.

Rajan, S D. "Sizing, Shape, and Topology Design Optimization of Trusses Using Genetic Algorithm", in: *Journal of Structural Engineering.* Volume 121, Number 10. pp. 1480. 1995

Rajeev, S. and Krishnamoorthy, C. S. "Discrete Optimization of Structures Using Genetic Algorithms", in: *Journal of Structural Engineering.* Volume 118, Number 5. p. 1233-1250. 1992.

Ramasamy, J. V. and Rajasekaran, S. "Artificial Neural Network and Genetic Algorithm for the Design Optimization of Industrial Roofs - a Comparison", in: *Computers and Structures.* Vol. 58, No. 4, pp. 747-55. UK. 17 Feb. 1996

Ramm, E., Bletzinger, K. U. and Maute, K. "Structural Optimization", in: Abel, J., Astudillo, R., and Srivastava (eds.) *Current and Emerging Technologies of Shell and Spatial Structures - Proceedings*, IASS/CEDEX, Madrid, 1997. pp. 201-216.

Research Engineers International. *STAAD.Pro2004.* Finite Element Analysis and Design Software. netGuru, Inc., 2004.

Rosenman, M. A. and Gero, J. S. "The Role of Functional Reasoning in Design". Key Center of Design Computing, Univ. of Australia, 1997

Russell, Peter J. *Genetics.* Harper Collins, New York, 1992.

Schank, Roger C. "I'm sorry, Dave, I'm afraid I can't do that", in: *Hal's Legacy*, The MIT Press, Cambridge, Mass., 1997. pp. 171-190.

Schildt, G. *Alvar Alto, the mature years.* Rizzoli, New York, 1989.

Schlaich, J. "On the Conceptual Design of Structures-an Introduction", in: *Conceptual Design of Structures*; Proceedings of the IASS International Symposium 1996, Stuttgart, Germany. Volume 1. pp. 15-25. Institut für Konstruktion und Entwurf II. Stuttgart, Germany, 1996.

Schmit, L. A. "Structural Design by Systematic Synthesis", in: *Proceedings, Second Conference on Electronic Computation*. ASCE, New York, 1960. pp. 105-122.

Schmitt, Heinrich. *Hochbaukonstruktion: die Bauteile und das Baugefüge Grundlagen des heutigen Bauens*, Otto Maier Verlag, Ravensburg, 1962.

Searle, J. R. "Minds, Brains, and Programs", in: *The Behavioral and Brain Sciences 3*, Cambridge University Press, 1980.

Seireg, A. A. and Rodriguez, J. *Optimizing the Shape of Mechanical Elements and Structures*. Marcel Dekker, Inc., New York, 1997.

Sekler, E. F. and Curtis W. *Le Corbusier at Work*. Harvard University Press, Cambridge, 1978.

Serrato-Combe, A. "The Wheel of Fortune and Pixel Farming", in: *Annual Meeting of the Association of Collegiate Schools of Architecture - Proceedings*, 1995.

Simon, Herbert A. *The Sciences of the Artificial*. 3rd ed. The MIT Press, Cambridge, Massachusetts. 1996. (1st ed. 1969)

Simon, Herbert A. "The Structure of Ill Structured Problems". *Artificial Intelligence*. Vol 4, Issue 3-4. North-Holland Publ. Co. 1973.

Simonton, Keith. *Origins of Genius: Darwinian Perspectives on Creativity*. Oxford University Press, 1999.

Slaby, Emanuel. *Einsatz Evolutionärer Algorithmen zur Optimierung im frühen Konstruktionsprozess*. Fortschritt-Berichte, Reihe 20. Nr. 361. VDI-Verlag, 2003.

Sobek, Werner. *Auf Pneumatisch Gestützten Schalungen Hergestellte Betonschalen*. Stuttgart Univ., Diss, 1987.

Soddu, Celestino. "Recoginability of the Idea: the evolutionary process of Argenia", in: *The AISB'99 Symposium on Creative Evolutionary Systems - Proceedings*, The Society for the Study of Artificial Intelligence and Simulation of Behavior, University of Edinburgh, 1999.

Stiny, G. "Ice-ray: a note on the generation of Chinese lattice designs", in: *Environment and Planning B: Planning and Design*, Vol. 4, 1977.

Stiny, G. and Mitchell W. J. "The grammar of paradise: on the generation of Mughul gardens", in: *Environment and Planning B*. Vol. 7. Pion Ltd., London. 1980. pp. 209-226.

Stiny, G. and Mitchell, W. J. "The Palladian grammar", in: *Environment and Planning B: Planning and Design*, Vol. 5, 1978.

Stipp, David. "2001 is just around the corner. Where's Hal?", in *Fortune.*, November 13., 1995.

Storrer, W. A. *The Frank Lloyd Wright Companion*. University of Chicago Press, Chicago, 1993.

Straub, H. *A History of Civil Engineering*. Leonard Hill, London, 1952.

Sugimoto, H. "Discrete Optimization of Truss Structures and Genetic Algorithms", in: *Structural Optimization*. Proceedings Korea-Japan Joint Seminar. Seoul Korea. pp. 1-10, 1992.

Syswerda, G. "Uniform crossover in genetic algorithms", in: *Proceedings of the Third International Conference on Genetic Algorithms*. Morgan Kaufmann Publ., 1989.

Tomlow, Jos. *The Model*. Karl Kraemer Verlag, Stuttgart, 1989.

Torroja, E. *The Philosophy of Structures*. University of California Press, Berkeley, 1967.

Turing, A. M. "Computing Machinery and Intelligence", in: *Mind* LIX, no. 2236, Oxford University Press, Oxford, 1950. pp. 433-460.

Venkayya, V. B. "Introduction: Historical Perspective and Future Directions", in: Kamat, Manohar P. (ed.) *Structural Optimization: Status and Promise*. Volume 150 Progress in Astronautics and Aeronautics. American Inst. of Aeronautics and Astronautics, 1993. pp. 1-10.

Vinacke, W. E. *The Philosophy of Thinking*. McGraw Hill, New York, 1953.

Wallas, G. *The Art of Thought*. Harcourt Brace, New York, 1926.

Watterson, Bill. The Essential Calvin and Hobbes: A Calvin and Hobbes Treasury. Andrews and McMeel (Universal Press Syndicate), New York, 1988.

Wenger, Win. *Discovering the Obvious*. Psychegenics Press, 1981.

Wertheimer, M. *Productive Thinking*. Harper, New York, 1945.

Wright, F. L. *An Autobiography*. Duell, Sloan and Pearce, New York, 1943.

Wright, F. L. *The Natural House*. Horizen Press, New York, 1954. pp. 182-183

Wu, Shyue-Jian and Chow, Pei-Tse. "Integrated discrete and configuration optimization of trusses using genetic algorithms", in: *Computers & structures*. Volume 55, Number 4. pp. 695, 1995.

Wu, Shyue-Jian and Chow, Pei-Tse. "Steady-state genetic algorithms for discrete optimization of trusses", in: *Computers & structures*. Volume 56, Number 6. pp. 979, 1995.

Xu, S. and Xia, R. "Topology Optimization of Truss Structures via the Genetic Algorithm", in: *Aerospace Technology and Science* - First Asian-Pacific Conference. International Academic Publishers, 1994.

Yang, J. and Soh, C.: "An Integrated Shape Optimization Approach Using Genetic Algorithms and Fuzzy Rule-based System". in: Topping, B.H.V. (ed.). Developments in Neural Networks and Evolutionary Computing for Civil and Structural Engineering. Civil-Comp Press, Edinburgh, UK, 1995.

Zarubin, V. A. "Genetic Algorithm in the Roll of a Shell for Structural Evolution Simulation at the Conceptual Design Stage", in: *1st Online Workshop on Soft Computing*, Aug.19-30, 1996.

Appendix A: Example Input/Output Files for the IGDT

Example Input Data File

```
Pony Truss Bridge -  4 panel - 3 Point Loads   June 2004
TYPE plane truss
UNIT   in   lb              >> only as labels
SUPPORTS 3                  >> support components
  1 fx      0.0      0.0    >> NOTE: NO 0 vertex to be given here
  1 fy      0.0      0.0    >> they are all reassigned one number lower
  2 fy    720.0      0.0    >> so 1 becomes Vertices[0]
STATIC_JOINTS 7             >> static components:
  2 fx    720.0      0.0    >> these are not moved by GS
  3 fx    180.0      0.0
  4 fx    540.0      0.0
  5 fx    360.0      0.0
  3 fy    180.0      0.0
  4 fy    540.0      0.0
  5 fy    360.0      0.0
FIXED_JOINTS     5          >> this is the number of joints supported or stat.
MINIMUM_JOINTS   7          >> this is the minimum number of joints needed for
INCIDENCES       4          >> this is the number of fixed members.
  1  1  3                   >> this is still a bit tricky
  2  3  5                   >> you have to enter lower number first
  3  4  5                   >> eg. 3 4   not   4 3
  4  2  4
SPAN   720.0                >> used to scale output graphics
CONSTANTS    3
  E  29000000.              >> UNITS are PSI
  FY     36000.             >> UNITS are PSI
  DENSITY   .28356          >> UNITS are LB/IN^2
PROPERTIES 1
  AREA    10                >> UNITS are IN^2, this is simply an initial estimate
COMBINATIONS_OF_LOADINGS  1 3  >> number of combs. and max number of loads
SELF_LOAD   fy   -1         >> self load for all combinations: downward
LOAD_COMB   1                        >>  the load comb id
  POINT    3   fy   -45000.0   >>  LBS downward on joint
  POINT    4   fy   -45000.0   >>  LBS downward on joint
  POINT    5   fy   -45000.0   >>  LBS downward on joint
FINISH
```

Example Geometry File for Progenitor

```
START OF TOPO DATA for Progenitor 'Pony'
JOINTS    10
MEMBERS   19
INCIDENCE:
1     1     3
2     1     6
3     2     4
4     2     10
5     3     5
6     3     6
7     3     7
8     3     8
9     4     5
10    4     8
11    4     9
12    4     10
13    5     7
14    5     8
15    5     9
16    6     7
17    7     8
18    8     9
19    9     10
COORDINANTS:
1     0.000000      0.000000
2     720.000000    0.000000
3     180.000000    0.000000
4     540.000000    0.000000
5     360.000000    0.000000
6     62.383455     128.996252
7     178.299621    203.824300
8     332.697695    247.000103
9     503.206900    222.903619
10    668.057358    125.033045
```

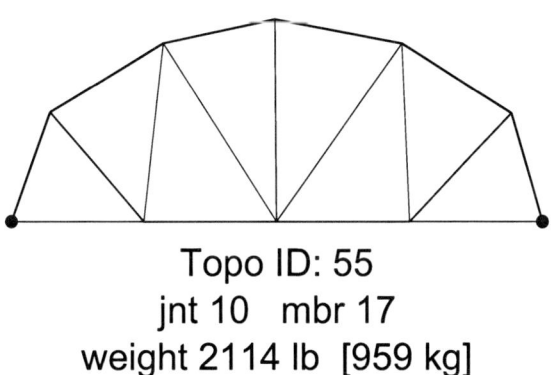

Topo ID: 55
jnt 10   mbr 17
weight 2114 lb [959 kg]

Figure A1. Plot of truss geometry matching the output text.

Example Output Text File for one Geometry

```
START OF TOPO DATA
NUMBER   55
JOINTS   10
MEMBERS  17
INCIDENCE:
1     1     3
2     1     6
3     2     4
4     2     10
5     3     5
6     3     6
7     3     7
8     4     5
9     4     9
10    4     10
11    5     7
12    5     8
13    5     9
14    6     7
15    7     8
16    8     9
17    9     10
COORDINATES:
1       0.000000     0.000000
2     720.000000     0.000000
3     180.000000     0.000000
4     540.000000     0.000000
5     360.000000     0.000000
6      53.000000   144.000000
7     206.707112   234.855042
8     358.006893   266.380434
9     529.710342   234.573769
10    678.538089   142.782330

Member Forces:
1       25183.261214
2      -72909.722960
3       19874.216846
4      -71268.013094
5       54902.755957
6       46677.972531
7       10227.705798
8       53367.046653
9       10197.155344
10      48738.763409
11      11863.203203
12      23951.729019
13      13987.864502
14     -65119.223850
15     -62705.411437
16     -62613.601796
17     -63226.122288
fitness    7455.487732
volume     7455.487732
weight     2114.078101
```

Appendix B: Graphic Depiction of the Geometry CHC

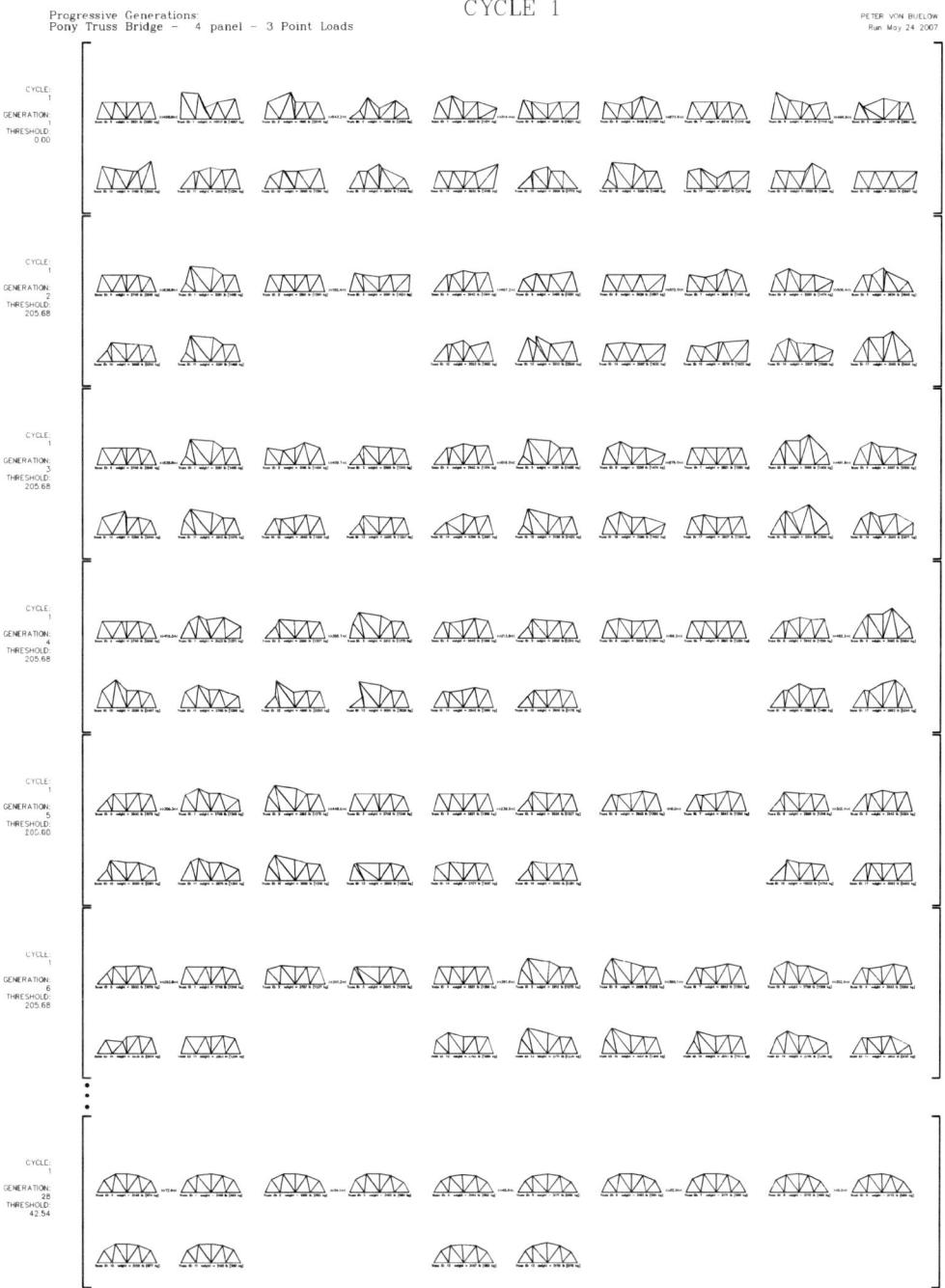

Figure A2. One cycle from the CHC geometry GA from example 3.1. (generations 1-6 + 28). For the purpose of illustration the parent population size is limited to 10 (normally 50).

Appendix C: Graphic Depiction of the Topology ES

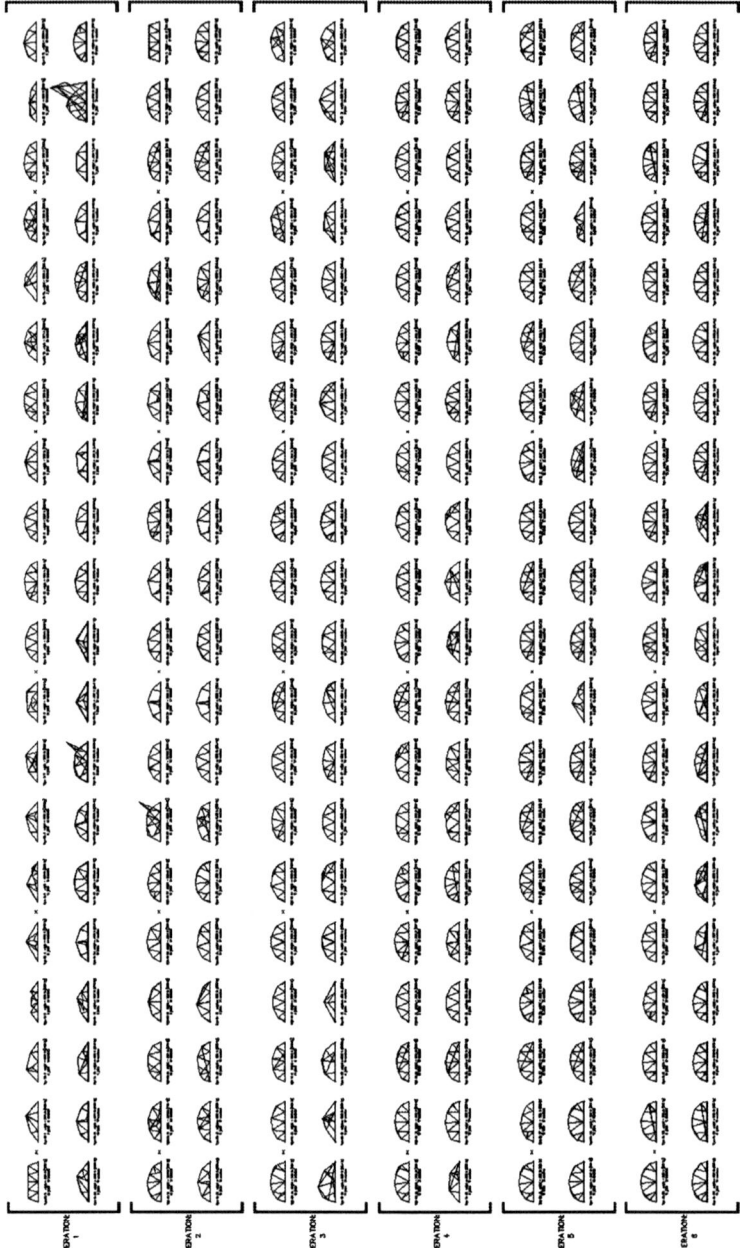

Figure A3. Six topology generations beginning with the progenitor used in example 3.1.

Appendix D: Analysis Assessment

This Appendix shows a comparison of the analysis results from the FEA code used in the IGDT with a commercial FEA code, STAAD.Pro 2005. The data from Topo 1, shown in Figure 3.3. of the pony truss example, is used in the comparison of results. Both programs use the 1989 edition of the AISC - ASD steel code for member analysis.

Two STAAD.Pro runs were made. The first run uses the same sizes chosen by the IGDT. The second run uses tabular sizes chosen by STAAD. There is a small discrepancy between the IGDT and STAAD in the member sectional properties since STAAD sections are 36 face polygons, while the IGDT calculated properties are based on circular sections. Also in rounding the steel density there is 0.2% difference between STAAD and the IGDT. This also has a small effect on the dead load and therefore member forces.

Finally, as the geometry was taken from a truss generated by the IGDT, it is not perfectly symmetric and therefore there a small differences in left and right sides.

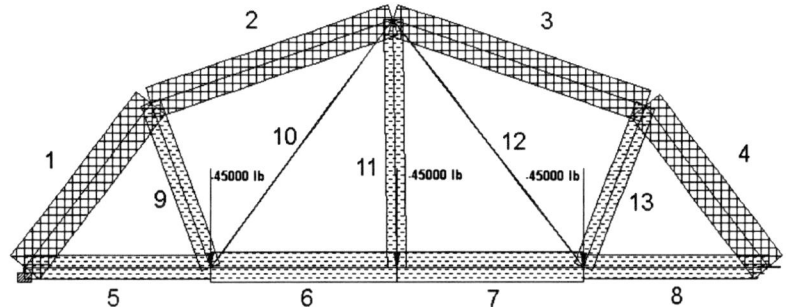

Figure A4. STAAD plot showing numbered members scaled to axial force.

Comparison of Member Forces

Table A2. shows a comparison of member forces found using STAAD and the IGDT. All results are within 2% (most much closer). As mentioned above there is some variation in the dead load which affects the member forces.

Comparison of Section Areas

Members in both STAAD and the IGDT were sized to the criteria of the 1989 AISC – ADS Steel Code. In the IGDT member sizes are chosen form an algorithm which relates a continuous range of sizes to wall thicknesses. The IGDT analysis does not use a slenderness (L/r) limit, but does limit minimum sectional area to 0.5 square inches.

STAAD.Pro uses a table of standard sizes. The smallest tabular size is ½ inch schedule 40 pipe with an area of 0.25 square inches. STAAD also uses the AISC slenderness limits of 200 for compression and 300 for tension.

Table A2. shows a comparison of the section areas. In the case of the IGDT the sizes are

MEMBER	IGDT	STAAD.Pro
1	-87333	-87500
2	-74527	-74600
3	-75585	-75600
4	-87511	-87600
5	54086	54100
6	70239	70200
7	70946	70900
8	54366	54400
9	47593	47600
10	786	803
11	45242	45300
12	1242	1260
13	47525	47600

Table A1. A comparison of the analysis results of the IGDT and STAAD.Pro 2004 showing member force in pounds.

chosen using a continuous equation to be at 100% capacity. Therefore, the ratio of actual stress / allowable stress = 1.00. Since STAAD picks standard sizes from a table the ratio is something less than 1.0. Factoring the areas by the capacity ratios gives results that compare closely.

Members 10 and 12 are long tensile members with very low load levels. In the IGDT they are sized by the minimum area criterion (0.50 sq. in.), and in STAAD they are sized by maximum slenderness ratios (300).

The total weight of the IGDT truss is 2586 LBS, and the STAAD.Pro truss totals 3498 LBS. When the IGDT member sizes were analyzed with STAAD they came within 0.05% of capacity.

MEMBER	IGDT x ratio	STAAD x ratio
1	6.09 x 1.00 = **6.09**	8.4 x 0.737 = **6.19**
2	6.28 x 1.00 = **6.28**	8.4 x 0.787 = **6.61**
3	6.55 x 1.00 = **6.55**	8.4 x 0.843 = **7.08**
4	6.04 x 1.00 = **6.04**	8.4 x 0.729 = **6.12**
5	2.50 x 1.00 = **2.50**	2.66 x 0.947 = **2.52**
6	3.25 x 1.00 = **3.25**	3.68 x 0.888 = **3.27**
7	3.28 x 1.00 = **3.28**	3.68 x 0.897 = **3.30**
8	2.52 x 1.00 = **2.52**	2.66 x 0.951 = **2.53**
9	2.20 x 1.00 = **2.20**	2.23 x 0.992 = **2.21**
10	0.50 x 1.00 = **0.50**	2.23 x 0.017 = **0.04**
11	2.09 x 1.00 = **2.09**	2.23 x 0.941 = **2.10**
12	0.05 x 1.00 = **0.05**	2.23 x 0.026 = **.058**
13	2.20 x 1.00 = **2.20**	2.23 x 0.990 = **2.21**

Table A2.   A comparison of the analysis results from the IGDT and STAAD.Pro 2004 showing member area in square inches.